Biodiversity, Ethnobotany and Ethnopharmacology

Biodiversity, Ethnobotany and Ethnopharmacology

– *Editors* –

Dr. A. Doss

Dr. V. Mary Kensa

Mr. N. Ahamed Kabir

Dr. G. Rajeswari

2026

Daya Publishing House®
A Division of
Astral International Pvt. Ltd.
New Delhi – 110 002

ISBN 978-93-5919-596-4

Publisher's Note:

Every possible effort has been made to ensure that the information contained in this book is accurate at the time of going to press, and the publisher and author cannot accept responsibility for any errors or omissions, however caused. No responsibility for loss or damage occasioned to any person acting, or refraining from action, as a result of the material in this publication can be accepted by the editor, the publisher or the author. The Publisher is not associated with any product or vendor mentioned in the book. The contents of this work are intended to further general scientific research, understanding and discussion only. Readers should consult with a specialist where appropriate.

Every effort has been made to trace the owners of copyright material used in this book, if any. The author and the publisher will be grateful for any omission brought to their notice for acknowledgement in the future editions of the book.

Published by : **Daya Publishing House®**
A Division of
Astral International Pvt. Ltd.
– ISO 9001:2015 Certified Company –
4736/23, Ground Floor, Ansari Road
Darya Ganj, New Delhi-110 002
Ph. 011-43549197, 8130496929
e-mail: info@astralint.com
Website: www.astralint.com

Printed at : **Replika Press Pvt Ltd**

Dr. P. Leema Rose
Assistant Professor
Department of Botany

Foreword

The relationship between nature and human well-being has been nurtured for centuries through traditional knowledge systems. Biodiversity, as the foundation of life on Earth, provides the raw material for human sustenance, while ethnobotany and ethnopharmacology bridge the wisdom of indigenous practices with modern scientific advancements.

This edited book offers an insightful exploration of the rich botanical heritage that has shaped traditional medicine and healing practices. It not only highlights the vast diversity of medicinal plants but also underscores the significance of preserving indigenous knowledge for sustainable healthcare solutions. Through a meticulous examination of plants and their pharmacological properties, the book serves as a valuable resource for researchers, conservationists, and practitioners alike.

In an era where biodiversity is threatened and traditional wisdom is at risk of being lost, this work stands as a crucial reminder of the need for conservation, ethical bioprospecting, and integrative approaches to medicine. May this book inspire deeper research, foster cross-disciplinary collaboration, and contribute to the sustainable use of nature's treasures for the benefit of humanity.

Sr. Leema Rose
Department of Botany,
Holy Cross College (Autonomous),
Nagercoil - 629 004.

Holy Cross College (Autonomous)
(Accredited with A++ Grade (CGPA 3.53-V Cycle) by NAAC)
Nagercoil, Kanyakumari District
Tamil Nadu, India - 629004

04652-261473
holycrossngl.edu.in

Dr. V. BALAKRISHNAN, Ph.D., F.S.A.B.,
Assistant Professor
PG and Research Department of Botany
Arignar Anna Government Arts College
Mohanur Road,Namakkal-637 002,
Tamil Nadu, INDIA

E Mail: palanivbalu@gmail.com

Mobile: +91- 94864 -85421

Biodiversity encompasses the various forms of life present in a specific region, including a wide range of animals, plants, fungi, and even microorganisms such as bacteria that comprise our natural environment. Each species and organism collaborates within ecosystems, functioning like a complex web to preserve equilibrium and sustain life. Biodiversity, which is a shorthand term for biological diversity, indicates the extensive variety of life on our planet, covering everything from genes to ecosystems and the relationships among them.

Conserving biodiversity means safeguarding and overseeing the variety of life on our planet, which encompasses species, ecosystems, and genetic diversity, to guarantee their enduring survival and the sustainable utilization of natural resources.

Biodiversity conservation involves safeguarding and managing biodiversity to secure resources for sustainable growth. The primary aims of biodiversity conservation are: to maintain species diversity; to ensure the sustainable use of species and ecosystems.

Ethnobotany is a multidisciplinary field that intersects natural and social sciences, examining the connections between humans and plants. It emphasizes the traditional understanding of how plants are utilized, cultivated, and regarded within human communities.Ethnobotany, which examines how communities utilize local plants, provides numerous benefits such as the identification of novel medicinal resources, a deeper appreciation of indigenous knowledge, and the encouragement of sustainable methods.

Pharmacognosy focuses on understanding natural drugs and their characteristics, whereas phytochemistry investigates the chemicals found in plants. Both disciplines play a crucial role in uncovering new medications and creating herbal products.

Pharmacognosy explores the physical, chemical, biochemical, and biological characteristics of drugs derived from natural sources. It encompasses the quest for new pharmaceuticals from these natural origins. The field involves the examination of various entities, including plants, animals, fungi, microbes, and minerals. It also covers the investigation of different parts of organisms, such as flowers, leaves, and roots. Additionally, it includes the study of drugs derived from microorganisms and animals. Notably, it involves the research of natural substances like morphine, atropine, and galanthamine.

The process of drug discovery involves finding and understanding molecules that can effectively and safely influence disease, aiming to develop medications that enhance patient well-being. This process is time-consuming and demands considerable resources, necessitating collaboration among various fields.

Place: Namakkal

Date:21.03.2025

Sincerely Yours

Preface

The world as we know it, is today at a cross roads. We are staring at what could turn out to be one of the mega extinction events in history. However, unlike the previous mega extinction events which had different reasons, this event is going to be the outcome of the activities of one species, us, *Homo sapiens*. We are also not sure whether the human species will be able to survive the mega extinction event it is helping to organise.

The planet is besieged on multiple fronts. From massive deforestation, to the use of xenobiotics and plastics and consequent dumping of plastics into our environment and ultimately into our oceans, to the release of carbon dioxide as a result of our activities which now has snowballed to global warming and climate change, the list seems endless. Man has so changed the world, in which he lives, that it is no longer possible to support the current lifestyle of the human race. Most probably it seems we are in for a difficult period in our history.

In this current scenario, what does this book on, "*Biodiversity, Ethnobotany and Ethanopharmacology*" have in store? Our current lifestyle and economies are characterised by a consumption mindset. The raw materials we need for achieving this mindset are taken from the natural environment. We tear apart rocks, cut forests, drill deep into the earth's mantle – all in search of energy and precious resources. Most of the materials that we consume such as metals and petrochemical products are non-renewable. Yet the race to acquire and consume more and more of the materials continues unabated.

Yet, could we think of another way forward? Could we source our raw materials for growth from the natural environment in a way that we nurture and protect our forests and natural resources? There are different thought processes and scientists and people are working in different ways towards that goal. Although their numbers are small at the present, it may lead us to a different paradigm of growth and development where the planet is not sacrificed at the altar of development.

There is currently research undergoing to construct the wing blades of the wind mills used for generating electricity using wood and it is expected that they may become a reality soon as prototypes are now being installed around the world. The African Green Belt initiative is another way of using this planet and the numerous plants and animals living on this planet to save the planet, to generate jobs and create livelihoods. Cultivation of sea algae along the seashores of the world by coastal communities is seen as another way to sequester the carbon dioxide in the air and also to provide the coastal communities with employment and an assured livelihood around the year.

We have to realise that the human species is a product of the environment and we are dependent on the environment for all our needs. Whether it is for food, for medicine or all our other needs, we have to depend on the natural environment. Nature's bounty is plentiful and there are still many more bio resources which we can develop for not only our needs but also to have a greener and healthier planet. What is needed is a change in the mindset of the human race and the leadership that will realize the dangers that the current model of development poses to our survival.

In this context, I hope that we can do something at our level as academicians to point out the various possibilities available in front of us which can be used to create a better world. I hope that the contributions by the various authors will pave the way in that direction. Together let each and every one of us do our little bit to save the planet.

Dr. A. Doss

Dr. V. Mary Kensa

Mr. N. Ahamed Kabir

Dr. G. Rajeswari

Contents

The Editors

Dr. A. Doss, PhD, is an Assistant Professor of Botany at V. O. Chidambaram College, Tuticorin, Tamil Nadu, India. He has Nine years of teaching experience and fifteen years of research experience in ethnopharmacology. He has received his graduation and post-graduation degrees from St. Joseph's College, Trichy, India, and a Doctor of Philosophy from National College, affiliated with Bharathidasan University, India. He has published more than 200 research papers in various international and national peer-reviewed refereed journals and seven books. His Google Scholar citation is 2538, with an H index of 27 and an I-10 index of 60.

Dr. V. Mary Kensa is working as an Associate Professor in S. T. Hindu College, Nagercoil. She is an active researcher with interests in the area of Pharmacognosy, Pharmacology and Phytochemistry in Botany. She has a teaching experience of 24 years having worked for 7 years in schools and the remainder in the college where she now works. Being an active academician, she has participated in a number of Orientation programmes and training programmes.

She is an active researcher and five of her students have been awarded their Ph.D. degrees under her guidance. At present four students are working under her to obtain their Ph.D. degrees. She has published 128 research papers in various national and international journals. She has written four books and contributed five research articles to various books. Her area of interest in writing books mainly deals with flowering plant diversity as reflected in the title of her books such as, Green Treasure of S. T. Hindu College, Nagercoil, Impact of pollution on *R.tuberosa* and *Hydrilla verticillata,* A Hand Book of Flowering plants and Fundamentals of Botany.

She has received a number of awards such as Best Researcher Award, Best Paper Presenter Award, Outstanding Faculty Award *etc.* a number of times. She regularly participates in seminars and conferences in India and abroad. She has been a participant of an international conference at AIMST University in Malaysia. She has also organized a number of seminars in her college and has served as a resource person for many conferences and seminars in the country. During the covid pandemic, she served as a resource person for many webinars.

As a person keenly involved in botany and the problems of biodiversity loss that we see currently, she has taken a special interest in documenting the biodiversity of her campus. She has also published a photo book of all the plants available in her college campus. Apart from developing her knowledge of plant taxonomy, she has also undertaken the plant audit of all the plants in the campus of St. Xavier's Catholic College of Engineering at Chunkankadai in Kanyakumari District.

She also has focused her research on solving the problems of taxonomy and diversity of many plants available in her native place and other places in Kanyakumari District. She has sequenced the gene sequences of *Heliotropium crispum* and submitted the same to the gene bank at NCBI. She is a member of many academic organizations the chief of which is the Society of Innovative Researchers and Academicians (SIRA) Kottayam District, Kerala, The Indian Science Congress Association, Kolkata, Society for Biotechnologists, India and Environment and Social Development Association, New Delhi. She is best described as a committed teacher who has the best interests of her students at her heart apart from being an academician and an active researcher.

Mr. N. Ahamed Kabir, M.Sc., is a Research Scholar of Botany at V. O. Chidambaram College in Tuticorin, Tamil Nadu. He obtained his undergraduate and post-graduate degrees from V. O. Chidambaram College in Thoothukudi, India. He has specific in researching the dynamic interactions between humans and plants using statistical and computer tools. Knowledgeable in fieldwork, data collecting, and advanced quantitative analysis to investigate traditional ecological knowledge, biodiversity protection, and sustainable resource management. Committed to developing a better knowledge of plant-human interactions in order to enhance cultural heritage and ecological resilience.

Dr. G. Rajeswari, Ph.D., is an Assistant Professor of Botany at Vellalar College for Women, Thindal Erode, Tamil Nadu, India, She has five years of teaching experience and years of research experience in as well as pharmacology. She has published 5 research papers related to medicinal plants in various National and International peer reviewed journals. Her Google scholar citation is 24 with an h-index of 15 and an i-10 index 12.

Chapter 1

Biodiversity Conservation and Management

Mary Kensa V.

Associate Professor, Department of Botany, S.T. Hindu College Nagercoil, Affiliated to MS University, Tirunelveli– 629002, Tamil Nadu
e-mail: surejkensa@gmail.com

ABSTRACT

India, one of the world's 12 mega-biodiversity nations, is separated into 10 biogeographical areas. Of the 35 biodiversity hotspots in the world, our nation is home to two: the Indo- Malayan region, which comprises the Eastern Himalayas, north-east India, and the Andaman Islands, and the Western Ghats. India's biogeographic location at the intersection of the Afrotropical, Indo-Malayan, and Paleo Arctic regions means that it possesses traits from all three. The combination of these three diverse realms gives the nation a rich and distinctive biological variety. It has a great wealth of biological diversity in its forests, wetlands and in its marine areas. India may have a great diversity of life, but there is seldom any opportunity for a sense of security. Threats to biodiversity include overuse of natural resources, climate change, changes in land use in natural areas, and the introduction of invasive species. Numerous initiatives, including a facilitation strategy and a formal framework, have been implemented to mainstream biodiversity and the environment. Conservation of biodiversity is protection, up liftment and scientific management of biodiversity so as to maintain it at its threshold level and derive sustainable benefits for the present and future generation.

***Keywords**: Biogeographic, Ecosystem, Generation, India, and Upliftment.*

1. What is Biodiversity

Biodiversity, the varieties of life on earth is the free gift of nature to planet earth and it is built over billions of years. Biodiversity is the foundation of life and it is important to human being for their sustenance, health, well being and recreation. It is a source of food, medicines, forest products, breeding stocks, population reservoirs and future resources for human being. It improves the life standards of human being through improved economy at local, national and international levels. Biodiversity provides social benefits such as recreation, culture and religious/ philosophical values. Biodiversity education provides opportunities for the better use of biodiversity for mankind. Biodiversity is unique, extremely complex, dynamic and varied and it unites the atmosphere (gases), geosphere (solid part) and hydrosphere (water) into one environment which offers scope for millions of species to live. Besides a source of food and medicine, it also performs several ecosystem functions. It helps in soil and water conservation, water storage, nutrient cycling and storage, pollution abatement and maintaining climate stability. Any damage or change in biodiversity will have serious impacts on humans and other living organisms on the earth. No other features of earth have been so dramatically influenced by man's activity. By altering biodiversity, we strongly affect human wellbeing and the wellbeing of every other living creature.

Distribution of biodiversity is important. Tropical countries are rich in biodiversity than temperate countries as tropical countries possess favorable climatic conditions for the growth of biodiversity. Tropical forests support diversity of fauna and flora. Forest litter that falls on soil surface enhances the production of microorganisms which facilitate the productivity of forest trees through mineralization process. The diversity of microorganisms in forest ecosystems is diverse which in association with soil and water produce nutrients, and a basic requirement for plant growth. Due to the dynamic nature of aquatic ecosystems such as freshwater, brackish water and sea water, both faunal and floral diversity are abundant in their respective ecosystems. The diversity is so unique in systems such as river mouths, coastal regions, wetlands, flood plain marshes and swamps as these environments receive huge quantities of nutrients of terrestrial origin which support rich biodiversity.

Biodiversity in aquatic ecosystems are rich and diverse and that the rich plankton production in near shore waters determines the productivity of commercially important fish and shellfish. Though production estimates are available, it cannot be completely relied upon as scientists are not fully aware of the distribution of picoplankton and its role in productivity estimates. This picoplankton is unicellular organisms with cell diameters of 1 and 4 microns and so far no standard methodologies have been developed to measure them.

Biodiversity perform different ecosystem functions which are essential for maintaining the ecological balance on earth. Although they perform the various ecosystem functions, so far it is not very clear how the diversity of genes, genotypes, species and communities influence ecosystem. Further, during the last 100 years a great volume of information are available on biodiversity and their direct and indirect values, a comprehensive, rigorous and general theory of biodiversity is lacking.

Ecosystem equilibrium is vital for various organisms to interact and perform their normal functions. As environment is on a constant state of transformation, ecosystem equilibrium cannot be balanced always, and this causes serious problems not only to biodiversity, but their vital functions. Balance of ecosystem is altered due to the fact that environment is subjected to different forces such as storms, floods, droughts, pests and parasites, out breaks of diseases, fires and human interference induced changes. According to reports, the diversity of life is facing serious threats and that, at the current rate of tropical forest loss, roughly 5to 10 per cent of forest species will be depleted per decade during the next twenty years. The extinction rate is restricted not only to forests, but it is also evident in aquatic ecosystems as these environments are under severe threat of natural and human induced factors. To understand climate change impacts, there is an imperative need to have detailed information on the biodiversity species availability and their wider distribution in varied ecosystems.

The values of biodiversity and to derive the maximum benefit out of it, it is necessary to have a systematic approach to identify the taxonomic position of biodiversity forms. Taxonomic research therefore becomes more vital and needs more attention as this will bring to light more and more species. Taxonomic research alone cannot solve many of the issues relating to biodiversity. The most important actions should be on collaborative initiatives to support biodiversity programmes. To achieve this, there is an urgent need to have collaborative actions of biologists, geneticists, physiologists, molecular biologists, ecologists and policy makers. National and international organizations have to collaborate to advance scientific understanding and importance of biodiversity. Public awareness on significance of biodiversity has to be created and this has to take deep roots among the society.

2. Concepts of Biodiversity

2.1. Meaning of Biodiversity?

Biodiversity refers to variety and variability among living organisms and the ecosystem complexes in which they occur. In a simplest sense, biodiversity may be defined as the sum total of species richness (number of plant species, animals and microorganisms occurring in a habitat).According to Global Biodiversity - Biodiversity is defined as the "totality of genes, species and ecosystems in a region. Convention on Biological Diversity signed by many member states at the earth summit held in Rio de Janeiro (Brazil) in 1992 explains biodiversity as: "*Biological diversity means the variability among living organisms from all sources including terrestrial, marine and other aquatic ecosystems and ecological complexes of which they are part; this includes diversity within species, between species and of ecosystems*".

Biodiversity comprises diversity of forms right from the molecular unit to the individual organism, and then on to population, community, ecosystem, landscape and biosphere levels. Biodiversity is described under three closely related aspects and they are genetic diversity, species diversity and ecosystem diversity. Broadly speaking, the term biodiversity includes three different but closely related aspects:

2.2. Genetic Diversity

This describes variation of genes within species. This describes variability among individual species. *e.g.* Each human being differs widely from others. This is because of large number of combinations possible in our genes that give specific characteristics. This genetic variability is essential for healthy breeding populations of a species. If the numbers of breeding individuals' are reduced, then there is inbreeding which leads to the extinction of the species.

2.3. Species Diversity

The number of species of plants and animals that are present in a region constitutes species biodiversity. Natural tropical forests have much greater biodiversity than most other regions. Modern agricultural systems have much fewer biodiversity than traditional agro pastoral farming systems where multiple crops are used. In intensive agriculture, only one type of crop is used but in traditional farming, farmers use many crops. Though intensive agriculture provides rapid economic returns, disease outbreaks can destroy the crop. In multiple crop system, if one crop fails, the other crop helps the farmer to compensate the loss. Natural forest has higher species diversity but in mono culture, only single species is cultured. In natural system, there are a large number of non wood products such as fruit, fuel wood, fodder, fiber, gum, resin, medicines *etc.* Timber plantations do not provide such goods for local consumption.

2.4. Ecosystem Diversity

Biodiversity is distributed in diverse ecosystems- Forests, grass lands, semi arid and desert, rangeland, aquatic ecosystems, islands, wetlands and swamps, each of which supports different and specific vegetation. Ecosystem diversity in contrast to genetic and specific diversity is difficult to measure since the boundaries of the communities are quite large with different communities, each community with definite species complexes.

2.5. Forests

The major factors that determine the biodiversity potential of forests are climate, soil and water. Forest encompasses a great variety of trees, shrubs and herbs and a wide variety of microorganisms in the habitat. The microbial population such as bacteria and fungi through mineralization process provide nutrients which form the food for plants. Forests are important because each forest has its own characteristic community structure with different life forms such as insects, reptiles, amphibians, birds and mammals. They also exhibit interrelationships such as mutualism, commensalism and parasitism.

2.6. Grassland

Grass lands are major sites where there is widespread cattle geazing. It is found in places where the rain fall is little which does not allow growth of large trees. Grass land supports the growth of different grasses and herbs. The biodiversity of grassland is unique with rich insect fauna. Birds are common in the region. Due to over grazing, grass land areas are now degraded due to compaction of soil by cattle. Further, the land degradation of grass land is due to the conversion of grass lands into agriculture land in many part of the country.

2.7. Semiarid and Desert Systems

Semiarid and desert are unique ecosystems as they are subjected to high temperature and scarcity of water, where evapo transpiration is higher than the precipitation. Due to this harsh climatically conditions, the biodiversity potential of this region is very much less and the region supports the growth of only the xerophytes plants.

2.8. Aquatic Ecosystems

Aquatic ecosystems are different kinds and dimensions. It is broadly classified as large oceans, brackish water bodies and freshwater systems. The freshwater bodies are of two types and they are lotic and lentic systems, each with its characteristic life forms. The biodiversity of aquatic ecosystems are rich and diverse and many of the life forms present in it are of great commercial significance. They provide food and medicine to large populations of the world. The forms that are of commercially significant include fish, crustaceans, mollusks, turtle, seaweeds and grasses.

2.9. Wetlands and Swamps

Wetlands are unique ecosystems found at the interface between land and water. They are productive ecosystems formed as a consequence of the interaction between the discharge of river water containing huge quantities of terrestrial nutrients and with the action of waves, tides and currents of the sea. Because of this interaction, this productive ecosystem supports diverse life forms. The region serves as good breeding and feeding grounds of many commercially important fish and crustaceans. Further they act as stop over for many of the migratory birds which come from distant countries and continents.

3. Distribution of Biodiversity

3.1. Global Biodiversity

Knowledge on biodiversity potential is crucial to make use of the biodiversity for human welfare and to maintain the ecosystem functioning. Hence availability of biodiversity data at national and international level becomes an urgent need. At present to take such decisions, the required data are not available and it is because, 1) Terrestrial and aquatic ecosystems are large and the biodiversity available in it are inexhaustible, collection of real data thus becomes challenging, time consuming and cost involved, (2).The global biodiversity data available are insufficient, scattered, outdated and mostly incomplete, (3) there is wide spread loss of biodiversity in the ecosystem due to multifarious factors. However, it is necessary to have comprehensive information on the available data for taking future course of action.

Different estimates are available with respect to global biodiversity. According to Gaston and Raven (1991) biodiversity number ranged between 4 to 40 million of which not more than 1.4 million species have been identified or named. Chapman (2005 and 2009) reports the biodiversity on earth is about 30 to 50 million, but only 1.9 million species have been identified or named (Source Chapman, 2009).

The species diversity is rich in tropics and more than half of the species on earth live in moist tropical forests which are only 7 per cent of the total land surface. In tropics species diversity was conserved over geological time and due to low rate of extinction prevailing there. Further, biological diversity is the result of interaction between climate, organisms, topography, parent soil materials, time and heredity.

3.1.1. Global Biodiversity Data

The United Nations Environment Programme (UNEP, 1993), suggested eight categories for the collection of biodiversity data for detailed study on biodiversity distribution and they are: biological, physical, socio-economic, cost benefit, biodiversity threats, Sustainable management, Sources and contacts, Interrelationships. Biological data deal with the information on species, genus and varied ecosystems. Physical data represent detail on climatically conditions, topography and hydrological regime of the area. Socio-economic data comprises socio economic status of the populations. Cost benefits confines to the value of biodiversity. Biodiversity threats represent the loss of biodiversity and the level of their threats. Sustainable management describes how the resources are utilized. Sources and contacts mainly deal with available literature, technologies developed and agencies and experts to be contacted for information on biodiversity. Interrelationships provide information on the relationships between and among species and ecosystems so as to forecast the effects of proposed actions.

Table 1.1: Mega Biodiversity Countries

Country	*Mammals*	*Birds*	*Flowering Plants*
Mexico	**450**	**1026**	**25000**
Indonesia	**436**	**1096**	**27500**
Zaire	**415**	**1096**	**11000**
Brazil	**394**	**1635**	**55000**

Country	Mammals	Birds	Flowering Plants
China	394	1224	30000
Columbia	359	1695	50000
Peru	344	1678	17121
India	316	1219	15000
Venezuela	305	1219	20000
Ecuador	302	1519	18250
Cameroon	297	0874	08000
Malaysia	286	0736	15000
Australia	252	0751	15000
South Africa	247	0790	23000
Panama	218	0926	90000
Papua New Guinea	214	0708	10000
Vietnam	213	0761	07000
Costa Rica	205	0850	11000
Phullippines	153	0556	08000
Medagascar	105	0253	09000

World ranking of mega-biodiversity countries (Pine, 1997).

3.2. Users of Biodiversity Data

National and international organizations are using the biodiversity for various purposes. The biodiversity users according to Crain and Reynolds (1999) are of three categories and they are National-level decision makers, International policy analysts, and the Biological scientific community. The governments, the general public, media, political parties, national NGO's, academies, scientific communities constitute the national institutions; regional groupings are ASEAN, MRC, OECD, G-7 countries while the international institutions are convention secretariats such as CBD,CITES, RAMSAR *etc.*; UN organizations such as UNEP, UNDP, UNESCO, FAO; international funding agencies such as WB, ADB, GEF *etc.*; bilateral development agencies namely DANIDA, NORAD, CIDA *etc.*; international environmental and conservation groups such as IUCN, WWF, Conservation International *etc.*

3.2.1. Criteria for Global Data Collection

Considering the significance of biodiversity in life and life management systems, efforts have been taken to acquire biodiversity data though various means. A number of global datasets have been acquired considering different criteria. The data on biogeography realms or provinces are prepared by combining physiographic and biogeography approaches (Udvardy, 1975). The structure of climax community has been used in identifying biogeography provinces at the first stage, then information on the presence of distinctive flora and fauna were used to delineate boundaries. This map divides terrestrial ecosystem of the earth into eight bio geographical realms that are further sub-divided into 193 provinces. Similarly Phyto geographic areas are delineated using floral information. Zoogeographic areas, on the other hand were delineated using faunal information. Cox and Moore (1993) based on physiognomy of the vegetation, provided the details on the distribution of biodiversity of the major terrestrial biomes of the world which include arctic tundra, northern coniferous forest, temperate forest, tropical rainforest, tropical seasonal forest, temperate grassland, tropical savanna, grassland and scrub, desert, Mediterranean vegetation and mountains. The Holdridge Life Zones of the World is divided according to a combination of climate and vegetation (ecological) types. The life zones were mapped based on climatic variables such as bio-temperature (based on the growing season length and temperature), mean annual precipitation, and a potential evapo transpiration ratio, linking bio temperature with annual precipitation to define humidity provinces. This map provides information on the generalized distribution of eco climatic zones across the globe, the underlying assumption of which is to expect similar types of vegetation under similar climatic conditions.

3.2.1.1. Species Data

Information on species is essential for documenting and describing biological diversity. This is supported by sound theoretical as well as operational reasons (Stanton and Lattin, 1989). As ecosystems are large and diverse it is extremely difficult to measure biodiversity on earth and that is the reason why exact number of biodiversity on earth is not known. Serious attempts are now in place to use alternative basis for measuring biological diversity. Different figures are now available on biodiversity estimates, however, a conservative estimate reveals that it ranges between 3 and 100 million species, and a total of 12.5 million species are known species and only 1.7 million species have been described (Table 1.2).

Table 1.2: Total Number of Species on Earth

Groups	*Known Number of Species*	*Estimated Total Number of Species*	*Per cent known Species*
Insects	950,000	8,000,000	11.9
Fungi	70,000	1,000,000	7.0
Arachnids	75000	750, 000	10.0
Nematodes	15000	500, 000	3.0
Viruses	5000	500, 000	1.0
Bacteria	4000	400, 000	1.0
Plants	250000	300, 000	83.3
Protozoan	40000	200, 000	10.0
Algae	40000	200, 000	10.0
Molluscs	70000	200, 000	35.0

Groups	Known Number of Species	Estimated Total Number of Species	Per cent known Species
Crustaceans	40000	150, 000	26.7
Vertebrates	45000	50, 000	90.0
World Total	1700000	12 500 000	13.6

WCMC, 1992.

Species richness can be determined based on the number of species occurring on an area. In reality it is not possible to get such details since because, the occurrence and composition changes according to time and space, and that many of the species are so minute identifying such organism are challenging. Further collecting and counting microorganisms from the environment is extremely difficult and it is not within the reach. The species diversity therefore is based largely on flowering plants, conifers and cycads, ferns and higher plants; vertebrate groups such as mammals, birds, reptiles, amphibians, freshwater fishes, and invertebrate groups such as butterflies and dragonflies. The data on flowering plants suggest that there are about 250,000 flowering plant species of which, 200 species have been domesticated as food plants of which only 25-30 are considered as crops of major world importance.

3.2.1.2. Genetic Data

Data and information on genetic diversity are available at three levels and they are: Genetic variability between individuals within population; Genetic variations among populations within species; and diversity among species. Genetic resources data are basically a repository of samples of living materials of animals, plants, fungi, or micro-organisms, generally in a dormant or other phase in which they are not actively growing (Hawksworth, 1995). Genetic resources information is collected mainly to ensure the conservation of genetic diversity in the future, the main purpose of which is to use it for breeding purposes. The collection can also be used as a source material for its characterization and evaluation. Thus, it is extremely important to conserve adequate stocks followed by regular monitoring and updating. The collections might include stores of gametes, pollen, seeds, spores, tissue cultures and embryos (Hawksworth, 1995). The storage can be maintained in long-term storage facilities as well as in field gene banks. Materials that are not possible to maintain in the dormant form can be maintained in field gene banks, as in the case of some cultivated crops.

3.3. Biological Reference Collection

These dataset are available in different data formats such as numeric, categorical (Classified or coded non- numeric data), text, spatial (map and GIS data), remote sensing (photographs and drawings), and sounds (*e.g.* voice of certain animal species).Biological reference collections comprise of permanently preserved specimens, recording and photographs maintained in museums, universities, botanical gardens, zoological and similar institutions or by individual scientists. Such collections are the major tools necessary for basic investigation and assessment of biodiversity.

Uses of Biodiversity

- ☆ Wide array of biodiversity is responsible for the existence of mankind
- ☆ He depends on them for food shelter, water and medicine
- ☆ Since ages man utilizes biodiversity

Biodiversity Classification

- ☆ Bioresources on earth can be classified as direct values and Indirect values
- ☆ Direct values are
- ☆ Biodiversity is a source of food
- ☆ Early days man consumed plants directly as food
- ☆ In due course of time wild species became the foundation of agriculture
- ☆ In traditional medicines, plants or their extracts are consumed directly or applied as medicines
- ☆ In modern medicine-199 pure chemicals from 90 plants
- ☆ Biodiversity is a source of antimicrobial, anti parasitic, cardio active and neurophysiologic substances
- ☆ Domesticate animals are sources of hormones, enzymes and food products
- ☆ Bacteria and fungi provides life saving medicines, such as antibiotics
- ☆ Biological resources from primary sources of wood, firewood and industrial products
- ☆ Ornamental plants are lucrative product today

Floral Biodiversity Uses

- ☆ Rice
- ☆ Rice crops are known as food
- ☆ Some varieties of rice posses medicinal properties (Chenellu)
- ☆ Some varieties of rice are flavored (Basumathi)

Tuber Crops

- ☆ Some tuber crops are used as food items
- ☆ Tapioca, Sweet Potato

Fruit Plants

- ☆ Fruits of many plants are consumed by man
- ☆ Plantains

- ☆ Mango
- ☆ Goosberry

Leafy Vegetables

- ☆ Several leafy vegetables are consumed by man
- ☆ Cabages, raddish

Spices

- ☆ A large number of spices are of importance as food
- ☆ Ginger
- ☆ Turmeric
- ☆ A number of them possess medicinal properties
- ☆ Ginger, turmeric, Pepper, cardamom, clove *etc.*,

Gums and Resins

- ☆ Cashew, Red cotton tree

Perfumes

- ☆ Perfumes are extracted from trees such as
- ☆ Sandal, Chambaka, Vetiver, lemon grass

Natural Dyes

- ☆ Some indigenous plants produce natural dyes such as
- ☆ Turmeric, red sandal wood

Lather

- ☆ Some plants are processed for the production of lather(Jaman)

Caner and Bamboo

- ☆ They form raw materials for paper industry
- ☆ Provide livelihood for rural population
- ☆ Household articles/handicrafts *etc.*

Oil

- ☆ Trees such as Neem are used for the production of medically important oils

Fibres

- ☆ Some plants species like coconut provide natural fibres

Fungi

- ☆ Fungi represent the most diverse groups of organisms
- ☆ Present in air, water soil and other organic materials

- ☆ Along bacteria they help in degrading biological material contributing to soil fertility
- ☆ Play a key role in environmental sanitation
- ☆ Its presence is inevitable for the germination of seeds such as that of Orchids
- ☆ Fungi associated with roots of plants help absorbing minerals and nutrients from the soil.
- ☆ Some fungi are delicious food items(Mushroom)
- ☆ Oyster mushroom and milky mushrooms are cultivated commercially
- ☆ Employed in the manufacture of bread, butter and beer
- ☆ They form raw material in the production of carbonic acid, alcohol, vitamins, weedicides, biological pesticides, antibacterial, antifungal drugs (Cyclosporin), used against the rejection of important organ by human body
- ☆ Fungal diseases are the major concern in agriculture, forestry and animal husbandry
- ☆ It may lead to great economic loss
- ☆ Bud Rot(Coconut), wilt(Peper), leaf blast(paddy), leaf spot disease (vegetables), fin and tail rot(Fish)
- ☆ Fungi responsible for 60 diseases in human beings and animals
- ☆ They are causative diseases for several skin diseases
- ☆ Degradation of wood is caused by fungi
- ☆ Fungi deteriorates the quality of many food items and render them inedible
- ☆ Wild life and their products serve mankind in several ways
- ☆ Fish and meat form important ingredients in our food
- ☆ Skin of snakes and crocodiles used for making articles(Purse, Kit *etc.*)
- ☆ Tusks of elephants, deer(Ivory products)
- ☆ Civet (catlike Animal in Central Africa), for preparation of drugs and perfumes
- ☆ Wild life watching- ecotourism, economy
- ☆ Animals are used in the testing the efficacy of drug
- ☆ Playing an important role in maintaining eco -balance(Snakes/rats; Birds and pollination)
- ☆ Human being depends on animal(food cloth, sports, entertainment, health *etc.*)
- ☆ Insect constitute the most dominating group in animal community
- ☆ Honey bee, silkworm, lack, insect, predator insect are beneficial to mankind)

- Agriculture and sericulture- cottage industry
- There are harmful insects(Malaria, Elephantiasis) by mosquitoes
- House fly carries pathogens
- Biological control by predatory insects reduce the application of chemical fertilizers and pesticides- thus reduce environmental pollution
- Biological control of pests and integrated pest management has become popular in agriculture and forestry
- Indirect values are:
- Carbon fixation through photosynthesis
- Pollination and gene flow *etc.*
- Maintaining water cycles, recharging ground water, protecting watersheds, control of floods and droughts
- Soil formation and protection from erosion
- Maintaining essential nutrient cycling
- Absorbing and decomposing pollutants
- Regulating climate at both macro and micro level
- Preserving recreational aesthetic, socio- cultural, scientific, educational, ethical and historical value of natural environment.

4. Medicinal Plant Biodiversity

Biodiversity is also precious "genetic library" maintained by several ecosystems. The potential of the genetic diversity of medicinal plants has not been tapped and thus their preservation and conservation is essential for economic benefit. Medicinal plant biodiversity is the variety of medicinal plants in the world, including their genetic diversity, species, and ecology. Medicinal plants are a vital source of raw materials for traditional herbal medicine and pharmaceutical drug discovery. They are used to produce vitamins, antibiotics, painkillers, and other medications. The distribution of medicinal plants varies by region. For example, China and India have the highest number of medicinal plants used, with 11,146 and 7,500 species, respectively. The biodiversity of medicinal plants is declining at an unprecedented rate due to human activities and environmental factors. For example, in India, the wild populations of about 100 of the traded species are known to have declined. To sustain the biodiversity of medicinal plants, it's important to understand the threats to them and their impact. One approach to conservation is to encourage farmers to participate in cultivating medicinal plants and conserving their natural habitats. This approach is based on the idea that local people and scientists can work together to achieve better results. Scientific advances are enabling the exploration of the medicinal value of plants and fungi. Emerging technologies can also help to safeguard essential medicines for the future.

Today, an estimated 80 per cent of people worldwide use traditional herbal medicines they are the first line of defense against disease for tens of millions of people. Despite the ubiquity and longstanding cultural significance of medicinal plants – totaling at least 30,000 species traditional uses of most medicinal taxa have not been evaluated clinically. In parallel, only 16 per cent of plants thought to have therapeutic value have been tested for biological activity.

Most contemporary research on medicinal plants is rooted in early- to mid-20th-century efforts that focused on the biochemical characterization, pharmacological activity, and synthesis of phyto molecules with potential medicinal utility. However, we suggest that the relationship between humans and medicinal plants may be more than one of identification, utilization, domestication, and harvest.

5. Rangeland Biodiversity

Rangelands are grasslands, shrub lands, woodlands, wetlands, and deserts that are grazed by domestic livestock or wild animals. Types of rangelands include tall grass and short grass prairies, desert grasslands and shrub lands, woodlands, savannas, chaparrals, steppes, and tundras. Rangelands do not include forests lacking grazable understory vegetation, barren desert, farmland, or land covered by solid rock, concrete, or glaciers.

Rangelands are distinguished from pasture lands because they grow primarily native vegetation rather than plants established by humans. Rangelands are also managed principally with practices such as managed livestock grazing and prescribed fire rather than more intensive agricultural practices of seeding, irrigation, and the use of fertilizers.

Grazing is an important use of rangelands but the term range land is not synonymous with grazing land. Livestock grazing can be used to manage rangelands by harvesting forage to produce livestock, changing plant composition, or reducing fuel loads.

Fire is also an essential regulator of range vegetation, whether set by humans or resulting from lightning. Fires tend to reduce the abundance of woody plants and promote herbaceous plants, including grasses, forbs, and grass-like plants. The suppression or reduction of periodic wildfires from desert shrub lands, savannas, or woodlands frequently invites the dominance of trees and shrubs to the near exclusion of grasses and forbs.

Rangelands cover approximately 80 million square kilometers globally, with 9.5 million square kilometers protected and 67 million square kilometers used for livestock production. These areas sustain about 1 billion animals, managed by pastoralists across over 100 countries, illustrating their crucial role in both ecological conservation and agricultural productivity. The United Nations (UN) has declared 2026 the International Year of Rangelands and Pastoralists, with the Food and Agriculture Organization leading the initiative.

Rangeland biodiversity is the variety of life and natural patterns in rangelands, which are geographical areas dominated by grasses and grass-like plants. Rangeland biodiversity is a key conservation target worldwide. It is largely determined by climate and disturbance regimes, such as grazing, fire, and extreme weather events. Rangelands are home to many large mammals and plants, and are important for a variety of reasons, including:

Economic Importance

Rangelands provide food and fiber, and forage for grazing animals.

Ecosystem Services

Rangelands help mitigate climate change, moderate weather, create and stabilize soil, and cycle nutrients.

Aesthetics

Rangelands can be visually interesting, and are popular for recreational activities like hiking and hunting.

Wildlife

Rangelands provide a diverse array of food sources, cover, and nesting sites for many species of wildlife.

Rangelands span a variety of ecosystems including grasslands, savannas, sagebrush steppe, shrub lands, tundra, mountain meadows and deserts. The variety of life and its processes on them (biodiversity) is important for moral, aesthetic and economic reasons, as well as for the services biodiversity provides to society.

5.1. Economic Benefits

Rangelands are vital economic assets, contributing substantially to national economies, particularly through livestock production. For instance, in Ethiopia, rangelands account for 19 per cent of the national GDP, while in Brazil, they contribute one-third of the agribusiness GDP through cattle farming. These vast areas not only support direct agricultural outputs but also bolster related industries, enhancing employment and promoting economic growth. Their management and sustainability are crucial for continuing these economic contributions and supporting the livelihoods dependent on them.

5.2. Rangeland Degradation Challenges

The degradation of Earth's extensive rangelands due to overuse, inappropriate cultivation, misuse, climate change, and biodiversity loss represents a significant threat to humanity's food supply and the well-being or survival of billions of people. In 2024, the United Nations Convention to Combat Desertification (UNCCD) reported that up to 50 per cent of rangelands are degraded. These areas suffer from reduced soil fertility, woody encroachment, erosion, salinization, alkalinization, and soil compaction, which all inhibit plant growth and contribute to drought and fluctuations in precipitation. This degradation is primarily driven by the conversion of pastures to cropland, urban expansion, increasing demands for food, fiber, and fuel, excessive grazing, abandonment by pastoralists, and policies that incentivize overexploitation. The UNCCD observes that the loss of rangeland attracts little public attention and rarely features in international policy discussions

6. Aquatic Biodiversity

Aquatic biodiversity is important because it helps maintain ecosystem services like water purification and climate regulation. It also provides resources like food and medicine for humans. Some threats to aquatic biodiversity include: increasing population and movements, a changing climate, coastal developments, and aquatic pests. Marine protected areas and other conservation strategies can help protect aquatic biodiversity. Aquatic biodiversity refers to the variety of life and ecosystems that exist in water, including the organisms that live in them:

6.1. Organisms

Aquatic biodiversity includes a wide range of organisms, such as fish, plants, algae, invertebrates, and microorganisms.

6.2. Ecosystems

Aquatic ecosystems include oceans, rivers, lakes, ponds, wetlands, estuaries, salt marshes, coral reefs, and mangroves.

6.3. Interactions

Aquatic biodiversity is comprehensive term that comprises freshwater ecosystems with lakes, ponds, reservoirs, rivers, streams, groundwater, and wetlands. The other part of aquatic biodiversity has marine ecosystems, which makes up an ocean, estuaries, salt marshes, coral reefs, mangroves and algal colonies. Different kinds of phytoplanktons, zooplanktons, aquatic plants, insects, fishes, birds and mammals are also an important part of aquatic biodiversity. Aquatic biodiversity also includes the interactions between these organisms and ecosystems.

Aquatic biodiversity is the rich and wonderful variety of plants and animals–from crayfish to catfish, from mussels to mayflies, from tadpoles to trout–that live in watery habitats. Many species of animals and plants live in water; some, like fish, spend all their lives underwater, whereas others, like toads and salamanders, may use surface waters only during the spring breeding season or as juveniles. Some aquatic creatures live their entire lives in the deep ocean, while others, like water striders, spend their lives skipping along the surface of water. Aquatic ecosystems (habitats and organisms) include our rivers and streams, ponds and lakes, oceans and bays, and swamps and marshes, and their associated animals. While freshwater ecosystems (excluding wetlands) cover only slightly more than two percent of the Earth's surface, they host almost 10 percent of the Earth's described animal species. The species richness of marine ecosystems is similar despite greater area. The southeastern US hosts a wide diversity of aquatic species including more than 1,200 fish species, nearly half of all salamander species in North America, almost half of crocodilian species, most of the world's temperate freshwater turtles, and some of the most diverse freshwater bivalve populations globally. Aquatic species have evolved and adapted to watery habitats over millions of years. Aquatic habitats provide the food, water, shelter, and space essential for the survival of aquatic animals and plants. Coastal estuaries and mangrove swamps, for example, are "edge" ecosystems that link salt- and freshwaters and trap nutrients that allow them to support a rich diversity of aquatic plants and animals.

Our aquatic wildlife are important sources of food, energy, jobs, atmospheric oxygen, buffers against new diseases, pests, and predators, and protection against food shortages and global climate change. Conserving a rich diversity of plants and animals will

- Help us to discover new drugs and medicines;
- Provide food for the growing human population;
- Help mitigate the impacts of climate change; and
- Add jobs and promote tourism through the enjoyment of nature.

Without the rich natural diversity of native plants and animals, our lives would be poorer, the supply of medicines more limited, career opportunities more scarce, and the economy less healthy. We depend on biodiversity just as much as we depend on clean water and air. We should make an exceptional effort to conserve wild species for tomorrow.

6.4. Threats to Aquatic Biodiversity

In December 2003, United Nations General Assembly, declared the year 2005 to 2015 as an international year for aquatic life, this official adoption was implemented to control the threats to aquatic biodiversity and their preservation and protection. The resolution demanded the serious concern on global water issues and development attempts, and nations were joined to fulfill the water related targets of the 2000 Millennium declaration. The threats to global aquatic biodiversity are very clearly known and the reasons can also be perceived; yet the extent of these damages varies from region to region. Identifying these threats can have little effect on the conservation, so it is advisable to mitigate the threats that can be a practical approach towards restoration. Colonization of new species in the aquatic system through human activities, whether knowingly or unknowingly, is becoming global environmental issue. The problem of introduced species is on the rise in United States freshwater resources. In the year 1920, six exotic fishes were introduced into the freshwater; a few more had been added by 1945 and until 1980 more species had been noticed. Newly arrived species in the ecosystem creates many problems for the native species, as they start breeding and population growth in more area other than their particular spot. Removal of chemical contaminations from the aquatic ecosystem can be eliminated at the point of source, or the habitat can be restored after pollution peak, but once the species introduction has begun, it is very difficult to reverse this task of introduction. Besides this, introduced or alien species can also produce negative impacts on water resources, such as preying of animals, competition, niche alteration, interspecific breeding and disease and parasitic infection development. Species are introduced to the new environment to gain economic advantage or to fulfill the developmental requirements like aquaculture. This introduction can be economically profitable initially but later it can produce disadvantages on economical and ecological terms, due to their invasive growth. Such an example is of Oysters, which has been introduced in to the global coastal waters for food. Oysters have function in estuarine ecosystem brings several complex modifications.

6.5. Conservation Approaches for the Aquatic Ecosystems

One of the remarkable approaches especially for the conservation of aquatic ecosystems was United Nations Convention on the Law of Sea (UNCLOS) which was implemented in the year 1994. This law was basically considering the promotion and support of the biodiversity along with the sustainable use of the oceanic water. Major non-government international organizations, research institutes and scientific groups can collaborate under the roof of the United Nations to work together for the assessment of global biodiversity and identification and listing of hotspots as biodiversity indicator. However, to prevent the future losses in the biodiversity either marine or freshwater, inadequate information availability should not be considered as hindrance in the progress of conservation. Communities operating at the broader sector should not delay waiting for the right time to get all the potential facts to initiate any process. Some of the data collected in the class of fishes and marine reptiles and other taxa are sufficient and reliable to trust that aquatic biodiversity is going through the crisis and can be calamity if not prevented.

Charles Elton, the "father of ecology", once presented his idea of conservation in his words that we should be looking for some wise principle of co-existence between man and nature, even if it has to be a modified kind of man and a modified kind of nature. Many wet land zones, whether fresh or brackish-water are biologically engineered by man or the environment is controlled by man. The best example is represented at Ramsar site, which hosts a large number of water birds which are under serious threats of extinction, another wetland in Indian subcontinent known as "Kabar Taal" is a winter niche for many migratory birds. Both these examples clearly show that human alterations in the ecosystem can be compatible and cooperative in biodiversity conservation.

Aquatic regions suffering from anthropogenic destruction, degradation or habitat loss should be aimed for the restoration. The same objective should imply for the conservation of the plant or animal species, which are facing decline growth and are close to extinction from the community. Water systems and aquatic biodiversity are closely related and interdependent on each other.

Bio-reserve spots can be created within a water body where fishing is restricted along with the other activities to protect the plants, animals and their habitat to conserve the aquatic biodiversity. Another useful role of these bio-reserves are that they can be highly beneficial for the purpose of education, recreation, tourism as well as potential increase in fishes reserve by improving the conditions of declining fish population. These bio-reserves function as marine protected areas, fishery reserves, sanctuaries and biological parks.

Bioregional management, can help to regulate factors influencing aquatic biodiversity, brings the balance in conservation, economy, and social demand of an area. This includes biosphere zones from small area and larger reserves.

Scientific reports have shown that habitats having extended biodiversity have chances to adapt in the new environment and re grows from various disasters either anthropogenic or natural. This can be considered beneficial in support of biodiversity since different species are performing the similar functions in a biologically diverse ecosystem; a disruption affecting one species may produce little impact on the entire ecosystem. Habitats with little diversity are considered as more vulnerable, as any insignificant interference in one species life may cause negative impact on the complex of interactions to collapse. Public awareness is demand of the time to teach citizens that only healthy functional aquatic ecosystems can provide all the benefits of improved water quality, water production and biodiversity richness. Immediate action in the form of strategic plans, economic incentives, public awareness and stakeholder involvement should be taken for the management and restoration of water resources and aquatic ecosystems. The management of water resources and aquatic ecosystems needs proper land management and sustainable implementation of land practices and holistic sense to identify the relevance between natural and man-made effects and developments. With the increase in population and human activities there will be more utilization of water resources so their biodiversity. Hence, restoration of aquatic habitat and conservation of biodiversity is the need of this modern time to maintain the quality of life.

7. References

1. Chapman, A. D. (2009). Numbers of Living Species in Australia and the World (PDF) (2nd ed.). Canberra: Australian Biological Resources Study. pp. 1–80.
2. Crain, J., and Reynolds, M. (1999). Utilizing high-throughput phenotypic data for improved phenotypic selection of stress-adaptive traits in wheat. *Crop Science*, 57(2), 648–659.
3. Udvardy, D. F. (1975) *World Biogeographical Provinces* (Map). The Co Evolution Quarterly, Sausalito, California.
4. Cox, C.B. and Moore, P.D. 1993: Biogeography: an ecological and evolutionary approach (5th edn). Oxford: Blackwell Scientific Publications.

5. Stanton, Nancy L.; Lattin, John D. 1989. In defense of species. BioScience. 30: 67.

6. WCMC (World Conservation Monitoring Center) (1992). Global Biodiversity: Status of the Earth's Living Resources (585 p.). London: Chapman and Hall.

7. D.L. Hawksworth.1995. The fungal dimension of biodiversity: magnitude, significance, and conservation. Mycological Research.

8. Angela L Duckworth, Patrick D Quinn, Eli Tsukayama.1993. What *No Child Left Behind* Leaves Behind: The Roles of IQ and Self-Control in Predicting Standardized Achievement Test Scores and Report Card Grades. J Educ Psychol. 2011 Dec 19;104 (2): 439–451.

9. Agarwal, N.K., 2011. Cryopreservation of Fish Semen In. J.P. Bhatt, Madhu Thapliyal and Ashish Thapliyal (eds.), Himalayan Aquatic Biodiversity Conservation and New Tools in Biotechnology, Transmedia Publication, Srinagar (Garhawal) Uttarakhand. pp: 104-127.

10. Agarwal, N.K., Raghuvanshi, S.K. and Saini, V., 2009. Cryopreservation of snowtrout (S.richardsonii) milt as a means for propagation and ex-situ conservation of species In. W.S. Lakra, A.K., Singh and P.C. Mahanta (eds.), Fish Genetic Resources, Narendra Publishing House, New Delhi. pp: 273-284.

11. Agarwal, N. K., Singh, G. and Rawat, U.S., 2014. Present status and threats to the Ichthyofaunal diversity of a snow fed river Nandakini in central Himalaya (Garhwal), India In.

12. Rawat U.S. and Semwal V.P. (eds.), Uttarakhand Disaster: Contemporary issue of Climate Change and Development with Holistic Approach, Winsar Publication, Dehradun, India. pp: 173-182.

13. Rawat, U., and Agarwal, N. (2015). Biodiversity: Concept, threats and conservation. *Environment Conservation Journal*, *16*(3), 19–28.

14. Agnihotri, N., Dassani, S., and T. K, S. (2020). Present Status and Conservation Strategies of Biodiversity in India. International Research Journal on Advanced Science Hub, 2, 251-255.

15. Antonelli, A., Kissling, W. D., Flantua, S. G., Bermúdez, M. A., Mulch, A., Muellner-Riehl, A. N., Kreft, H., Linder, H. P., Badgley, C., Fjeldså, J. and Fritz, S. A. (2018). Geological and climatic influences on mountain biodiversity. Nature Geoscience, 11(10), 718-725.

16. Bawa, K. S., Sengupta, A., Chavan, V., Chellam, R., Ganesan, R., Krishnaswamy, J., athur, V. B., Nawn, N., Olsson, S. B., Pandit, N. and Quad- er, S. (2021). Securing biodiversity, securing our future: A national Mission on biodiversity and human well-being for India. Biological Conservation, 253, 108867.

17. Chandrakar, A. K. (2012). "Biodiversity conservation in India, Technical Report" DOI: 10.13140/RG.2.1.1490.3208.

18. Dar, M. A. (2020). Biodiversity conservation in Jammu and Kashmir state: legal framework and concerns. In Biodiversity of the Himalaya: Jammu and Kashmir State. Springer, Singapore, 1077- 1095.
19. Elizabeth Dowdeswell (1995). In Global Biodiversity Assessement- UNEP, CUP, UK.
20. Ghosh, A. K. (2017). "Biodiversity, Access and Benefit Sharing in India: A Critical Analysis." In Biodiversity for Sustainable Development, pp. 79-86.
21. Springer, Cham. Java, R.L. and Srivastava Asheem (1998). Biodiversity conservation strategies for the 21st century, Biodiversity Conservation. 18- 31.
22. McNeely, J. A. (1995). Conserving the world's biological diversity. In: Global Biodiversity Assessment, UNEP, CUP, UK.
23. Meenakumari, B., and Rana, R. S. (2017). Implementing the Biological Diversity Act in India: Focus on ABS and Agro- biodiversity. Indian Journal of Plant Genetic Resources, 30(1), 25-30.
24. Onial, M., Jasmine, B., Singh, Y., Pande, A., Ramesh, C., Sivakumar, K., and Mathur, V. B. (2018). Updating India's National Biodiversity Action Plan: the process and way forward. Current Science, 115(3), 42.

Chapter 2

Biodiversity and Its Conservation

V. Vidhya

Assistant Professor, Department of Zoology, St.Jude's College, Thoothoor–629176, Kanyakumari District, Tamil Nadu
(Affliated to Manonmaniam Sundaranar University, Tirunelveli–627012)
e-mail: vidhyavvnair@gmail.com

ABSTRACT

Biodiversity conservation involves the protection, management, and restoration of biodiversity. Biodiversity is a measure of variation at the genetic, species, and ecosystem levels. People depend on biodiversity in their daily lives, in ways that are not always apparent or appreciated. Human health ultimately depends upon ecosystem products and services (such as availability of fresh water, food and fuel sources) which are requisite for good human health and productive livelihoods. Biodiversity harm can have significant direct human health impacts if ecosystem services are no longer adequate to meet social needs. Considering above all, this chapter discuss about the biodiversity conservation, methods of conservation endangered species and why it should be conserved.

Keywords: *Biodiversity, Consideration, Endangered species, Environment, Impacts.*

1. Introduction

Biodiversity is the summarized form of two words "biological" and "diversity". It reveals to all the variability of life that can be create on Earth (plants, animals, micro-organisms, *etc.*) as well as to the communities that they form and the habitats in which they live. Biodiversity can be also explained as the total of every plants and animals found in a particular area. It is primarily influenced by the environmental, geographical, climatic and topographical surroundings of an area. The word 'mega' represent 'large' therefore; mega diversity means a great number of species are found in the ecosystem. India is wealthy in the diversity of animals and plants, so India is called mega diversity region, especially India is one of 17 mega-biodiversity countries in the world. With only 2.4 percent of the earth's ground area, it accounts for 7-8 percent of the world's recorded species. One of seventeen mega diverse countries, it is address to 7.6 per cent for mammalian, 12.6 per cent for avian, 6.2 per cent for reptilian, 4.4 per cent for amphibian, 11.7 per cent for fish and 6.0 per cent for flowering plants. Forests are the most biodiversity-amusing earthly ecosystem, it supply a wide range of values to humans, varying from timber, pulp and rubber, to environmental services. At the worldwide level, forests play a critical part in regulating the climate and represent an important carbon reservoir.

The father of biodiversity is Edward O. Wilson, he was an entomologist and Wilson was faculty of Harvard University. In 1980, Thomas Lovejoy introduced the term biological diversity to the scientific community in a book. It rapidly became commonly used. The National Biodiversity Authority (NBA) is a statutory autonomous body under the Ministry of Environment, Forests and climate change, Government of India established in 2003 to implement the provisions under the Biological Diversity Act, 2002, after India signed Convention on Biological Diversity (CBD) in 1992. Charles Darwin identified the first principle of the origin of modern biodiversity, namely that all species were linked in a single great phylogeny, or tree of life and that all could be traced back to a presumed particular original species at some distant time in the geological past.

Biodiversity conservation involves the protection, management, and restoration of biodiversity. Biodiversity is a measure of variation at the genetic, species, and ecosystem levels. Biodiversity can be conserved by various methods including in situ and ex situ conservation methods that include establishing national parks, wildlife sanctuaries, *etc.* Biodiversity should be conserved as it provides us with both economic and aesthetic benefits. Considering above all, this chapter discuss about the biodiversity conservation, methods of conservation endangered species and why it should be conserved. Biodiversity conservation is the protection and management of wildlife and its habitats. It includes preserving and restoring species and their environments.

2. Importance of Biodiversity

Biodiversity is intrinsically valuable and is important for our emotional, psychological and spiritual well-being. Some consider that it is an important human responsibility to be stewards for the rest of the world's living organisms. Diversity breeds diversity. Having a diverse array of living organisms allows other organisms to take advantage of the resources provided. For example, trees provide habitat and nutrients for birds, insects, other plants and animals, fungi and microbes. Humans have always depended on the Earth's biodiversity for food, shelter and health.

Biological resources that provide goods for human use include:

i) Biodiversity Safeguards Food Safety

Biodiversity provides food from crops, livestock, fisheries, and forests. Biodiversity is the basis of agriculture and food systems. Agricultural biodiversity also performs ecosystem services such as soil and water conservation, maintenance of soil fertility, conservation of biota and pollination of plants, all of which are essential for food production and for human survival. It plays a vital role in human nutrition through its influence on world food production, as it ensures the sustainable productivity of soils and provides the genetic resources for all crops, livestock and aquatic species harvested for food. Biodiversity strengthens the global nutrition and food security.

ii) Biodiversity Protect Health and Stop Diseases

Human activities are disturbing both the structure and functions of ecosystems and altering native biodiversity. Patterns of infectious diseases are sensitive to these disturbances. Major processes affecting infectious disease reservoirs and transmission include, deforestation; land-use change; water management *e.g.* through dam construction, irrigation, uncontrolled urbanization or urban sprawl; resistance to pesticide chemicals used to control certain disease vectors; climate variability and change; migration and international travel and trade; and the accidental or intentional human introduction of pathogens.

iii) Biodiversity Provide Livelihoods

Biodiversity plays an important role in ecosystem functions that provide supporting, provisioning, regulating and cultural services. These services are essential for human well-being. Biodiversity provides building blocks for sustainable food, health and livelihood security systems. It is the feedstock for the biotechnology industry and a climate resilient farming system. In spite of the importance given to biodiversity conservation, genetic erosion continues globally.

iv) Biodiversity Prevent Air, Water and Soil Contamination

Biodiversity is the key indicator of the health of an ecosystem. Healthy ecosystems clean water, purify air, maintain soil, regulate the climate, recycle nutrients and provide us with food. A wide variety of species will cope better with threats than a limited number of them in large populations.

Biodiversity can play an important function in the cleaning of water. When water passes through lakes, wetlands, and streams, it often encounters different species of fungi, algae, and bacteria. Wetlands rich with these filtering organisms act as natural water filtration systems. Many of the microbes actually filter microscopic particles out of the water, making it safe for humans to drink. Even some larger species do similar work. For example, the caddisfly constructs nets underwater that filter out different kinds of particles, which it then eats. Wetlands rich with these underwater organisms act as natural water filtration systems. When the biodiversity of a region declines, many of the organisms critical to this filtering process can disappear. Therefore, pressures on the freshwater supply can cause biodiversity to decrease, which can cut the drinkable water supple even further.

Soil biodiversity reflects the mix of living organisms in the soil. These organisms improve the entry and storage of water, resistance to erosion, plant nutrition, and break down of organic matter. Soil pollution causes a chain reaction. It alters soils' biodiversity, reduces soil organic matter and soils' capacity to act as a filter. It also contaminates the water stored in the soil and groundwater and causes an imbalance of soil nutrients. Soil protection by maintenance of biological diversity can preserve the productive capacity of the soil, prevent landslides, safeguard coastlines and riverbanks, and prevent the degradation of coral reefs and coastal fisheries by siltation. Trees and other vegetation also assist in soil formation.

v) Biodiversity Contribute to Climate Strength

Biodiversity contribution to climate stability. Vegetation influences the climate at macro and micro levels. On smaller scales, vegetation has a moderate influence on local climates and may create quite specific microclimates. Some organisms are dependent on such microclimates for their existence. The link between climate change and biodiversity has long been established. Although throughout Earth's history the climate has always changed with ecosystems and species coming and going, rapid climate change affects ecosystems and species ability to adapt and so biodiversity loss increases.

vi) Maintenance of Ecosystems

Healthy ecosystems will be more resilient to climate change and so more able to maintain the supply of ecosystem services on which our prosperity and wellbeing depend. Biodiversity also provides supporting ecosystem services, which are necessary for the production of all other more direct ecosystem services. For example, by influencing primary production and nutrient and water cycling, biodiversity indirectly supports the production of food, fiber and shelter.

vii. Medicine

Biodiversity provides medicine from wild plants like Cinchona and Foxglove.

viii. Economy

Biodiversity supports industries like tourism, agriculture, fisheries, and forestry.

ix. Climate Change

Biodiversity helps mitigate climate change by sequestering carbon dioxide.

x. Ecosystem Resilience

Biodiversity helps ecosystems adapt to disturbances like disease outbreaks, natural disasters, and climate variability.

xi. Culture

Biodiversity is an important part of culture and identity.

xii. Inspiration

Biodiversity inspires and contributes to the beauty of natural landscapes.

3. Types of Biodiversity

Generally three types of biodiversity are as follows, they are species diversity, genetic diversity and ecosystem diversity.

i) Species Diversity

Species diversity is all the differences within and between populations of species, as well as between different species. Species Diversity is simply the number and relative abundance of species found in a given biological organization (population, ecosystem and Earth). Species diversity is a benchmark that can be used to evaluate the health of ecosystems. In a healthy ecosystem, a diverse and balanced number of species exist and maintain the equilibrium of the ecosystem.

ii) Genetic Diversity

Genetic diversity is all the different genes contained in all individual plants, animals, fungi and microorganisms. It happens within a species as well as between species. It also refers to the range of different inherited traits within a species. In a species with high genetic diversity, there would be many individuals with a wide variety of different traits. Genetic diversity is critical for a population to adjust to changing environments. There are three sources of genetic variation such as mutation, gene flow and sexual reproduction.

iii) Ecosystem Diversity

Ecosystem diversity is all the different habitats, biological communities and ecological processes, as well as variation within individual ecosystems. Ecosystem diversity deals with the variations in ecosystems within a geographical location and its overall impact on human existence and the environment. Ecosystem diversity addresses the combined characteristics of biotic properties (biodiversity) and abiotic properties (geodiversity).

4. Major Biodiversity Hotspots in India

India has four biodiversity hotspots, *i.e.*, Himalayas (Eastern and Western Himalayas), Western Ghats, Indo-Burma region and Andaman and Nicobar Islands. These hotspots have numerous endemic species.

i) Eastern and Western Himalayas

Eastern Himalayas form a part of the Himalayan global biodiversity hotspot. This area is exceptionally rich in diversity and endemism. It comprises of parts of Nepal, Bhutan, Sikkim, Arunanchal Pradesh and extends up to Burma. This region has an estimated 9000 plant species, out of which 3500 are endemic. Anthropogenic activities including deforestation, fragmentation of habitats, pollution, high population, climate change and poaching of wildlife poses a serious threat to the biodiversity of the region.

The Eastern Himalayas boast one of the world's richest counts of alpine flora within its temperate broadleaved forests, with a total of 10,000 species of plants. Out of these, 71 genera with 3,160 different species are endemic to the region, including a great number of species from the genus Rhododendron that can be found in great quantities on the hills of the Eastern Himalayas. With at least 7,500 flowering plants in India, 50 per cent of them are from this region, described as the "cradle of flowering plants". There are about 700 species of orchids in the north-eastern region of India alone. The Orchidaceae is the largest family, with 750 endemic species, out of which 545 (within 122 genera) are restricted to Arunachal Pradesh region, with 12 being endangered, 16 vulnerable and 31 near threatened. Eastern Himalayan region is home to 300 mammalian species, 977 avian species, 176 reptilian species, 105 amphibian species and 269 species of freshwater fishes.

Western Himalayas, also called Punjab Himalayas, westernmost section of the huge Himalayas mountain range. It lies mainly in the disputed Kashmir region of the northern Indian subcontinent including portions administered by India and Pakistan and also in the north-western part of Himachal Pradesh state, India. The Western Ghats are one of the world's biodiversity hotspots with over 5,000 flowering plants, 139 mammals, 508 birds and 179 amphibian species. At least 325 globally threatened species occur here. The Western Ghats of India are very rich in amphibian species with 117 species of frogs, toads and caecilians. About 89 species are endemic to this biogeographical region.

ii) Western Ghats

The Western Ghats, known locally as the Sahyadri Hills, are formed by the Malabar Plains and the chain of mountains running parallel to India's western coast, about 30 to 50 kilometers inland. They cover an area of about 160,000 km^2 and stretch for 1,600 km from the country's southern tip to Gujarat in the north, interrupted only by the 30 km Palakkad Gap. Sri Lanka is a continental island separated from southern India by the 20-meter-deep Palk Strait. The island, some 67,654 km^2 in size, has been repeatedly connected with India between successive inter-glacial, most recently until about 7,000 years ago by a land bridge up to about 140 kilometers wide. The Western Ghats mediates the rainfall regime of peninsular India by intercepting the southwestern monsoon winds. The western slopes of the mountains experience heavy annual rainfall (with 80 per cent of it falling during the southwest monsoon from June to September), while the eastern slopes are drier; rainfall also decreases from south to north. Dozens of rivers originate in these mountains, including the peninsula's three major eastward-flowing rivers. Thus, they are important sources of drinking water, irrigation, and power. The ecosystem in India's Western Ghats is severely endangered due to the increasing human settlements, mining, pollution and the drop in genetic diversity.

iii) Indo-Burma Region

The Indo-Burma region ranks among the world's top 10 biodiversity "hotspots", a term which is used to describe the planet's most biologically rich and threatened regions. The Indo-Burma Hotspot includes all non-marine parts of Cambodia, Lao PDR, Myanmar, Thailand and Vietnam, plus parts of southern China. Although this region is quite rich in its biodiversity, it has been worsening over the past few decades. Six species of mammals have been discovered in this region recently including large-antlered muntjac, AnnamiteMuntjac, grey-shanked douc, leaf deer, saola and Annamite striped rabbit. Other species such as monkeys, langurs, and gibbons can be found here with a population of less than a hundred. Freshwater turtle species found in the region are however endemic. 1300 species of birds too can be spotted here including the white-eared night-heron, Gray-crowned crocias, and orange necked Patridge most of which are endangered. Almost 13,500 plant species can be spotted in the region half of which are endemic and cannot be found in any other place in the world.

iv) Andaman and Nicobar Islands

This region lies in South-East Asia and includes Thailand, Singapore, Indonesia, Brunei and Malaysia. The Nicobar Islands represent India. These islands were declared as the world biosphere reserve in 2013 by the United Nations. These islands have a rich terrestrial as well as marine ecosystem including mangroves, seagrass beds and coral reefs. Species such as dolphins, whales, turtles, crocodiles, fishes, prawns, lobsters and seashells comprise marine biodiversity. In case the marine resources are over-used, it can pose a serious threat to biodiversity. Sundaland holds about 25,000 species of vascular plants, 15,000 of which are found nowhere else. There are at least 117 endemic plant genera in the hotspot; 59 of these endemic genera are found in Borneo, 17 in Sumatra, and 41 on the Malay Peninsula. Borneo boasts a spectacular diversity of trees. Sundaland is known for its significant biodiversity in the plant and animal species found in the area. Plant species include the Rafflesia, dipterocarps and orchids.

5. List of Endangered Animals in India

India's Red List of 2018 was released at the Rio+20 Earth Summit. Since then, new animals have been added yearly. While previously this list contained 132 species of plants and animals in 2018, as of the 2023-1 update from the IUCN Red List, over 950 species of animals (and over 600 species of plants) are listed as critically endangered, endangered, or vulnerable.

I. Critically Endangered Animals

Listed by the IUCN as of the 2023-1 update.

1. Arachnids

- ☆ Rameshwaram parachute spider (*Poecilotheria hanumavilasumica*)
- ☆ Peacock tarantula (*Poecilotheria metallia*)

2. Birds

- ☆ Great Indian bustard *(Ardeotis nigriceps)*
- ☆ White-bellied heron *(Ardea insignis)*
- ☆ Baer's pochard *(Aythya baeri)*
- ☆ Spoon-billed sandpiper *(Calidris pygmaea)*
- ☆ Yellow-breasted bunting *(Emberiza aureola)*
- ☆ Christmas Island frigatebird *(Fregata andrewsi)*
- ☆ Siberian crane *(Grus leucogeranus)*
- ☆ White-rumped vulture *(Gyps bengalensis)*
- ☆ Indian vulture *(Gyps indicus)*
- ☆ Slender-billed vulture *(Gyps tenuirostris)*
- ☆ Masked finfoot *(Heliopais personatus)*
- ☆ Bengal florican *(Houbaropsis bengalensis)*
- ☆ Bugun liocichla *(Liocichla bugunorum)*
- ☆ Himalayan quail *(Ophrysia superciliosa)*
- ☆ Jerdon's courser *(Rhinoptilus bitorquatus)*
- ☆ Pink-headed duck *(Rhodonessa caryophyllacea)*
- ☆ Red-headed vulture *(Sarcogyps calvus)*
- ☆ Lesser florican *(Sypheotides indicus)*
- ☆ Sociable lapwing *(Vanellus gregarius)*

3. Cnidaria

- ☆ Millepora boschmai possibly extinct

4. Fish

- ☆ Stripenose guitarfish *(Acroteriobatus variegatus)*
- ☆ Knifetooth sawfish *(Anoxypristis cuspidata)*
- ☆ Wayanad mahseer *(Barbodes wynaadensis)*
- ☆ Pondicherry shark *(Carcharhinus hemiodon)*
- ☆ Oceanic whitetip shark *(Carcharhinus longimanus)*
- ☆ Sand tiger shark *(Carcharias taurus)*
- ☆ Dwarf gulper shark *(Centrophorus atromarginatus)*
- ☆ Indian swellshark *(Cephaloscyllium silasi)*
- ☆ Sharpnose guitarfish *(Glaucostegus granulatus)*
- ☆ Halavi guitarfish *(Glaucostegus halavi)*
- ☆ Widenose guitarfish *(Glaucostegus obtusus)*
- ☆ Clubnose guitarfish *(Glaucostegus thouin)*

- Giant guitarfish *(Glaucostegus typus)*
- Ganges shark *(Glyphis gangeticus)*
- Kudremukh glyptothorax *(Glyptothorax kudremukhensis)*
- Tentacled butterfly ray *(Gymnura tentaculata)*
- Nilgiri mystus *(Hemibagrus punctatus) possibly extinct*
- *Horalabiosa arunachalami*
- Haragi *(Hypselobarbus pulchellus)*
- Red Canarese barb *(Hypselobarbus thomassi)*
- Pakistan whipray *(Maculabatis arabica)*
- Shorttail whipray *(Maculabatis bineeshi)*
- *Mesonoemacheilus herrei*
- Bovany barb *(Neolissochilus bovanicus)*
- Deolali minnow *(Parapsilorhynchus prateri) possibly extinct*
- Pookode Lake barb *(Pethia pookodensis)*
- *Pinniwallago kanpurensis possibly extinct*
- Dwarf sawfish *(Pristis clavata)*
- Common sawfish *(Pristis pristis)*
- Largetooth sawfish *(Pristis microdon)*
- Longcomb sawfish *(Pristis zijsron)*
- *Psilorhynchus tenura*
- Deccan barb *(Puntius deccanensis) possibly extinct*
- *Puntius madhusoodani*
- *Schistura papulifera*
- Bowmouth guitarfish *(Rhina ancylostoma)*
- Bengal guitarfish *(Rhinobatos annandalei)*
- Smoothback guitarfish *(Rhinobatos lionotus)*
- Bottlenose wedgefish *(Rhynchobatus australiae)*
- Smoothnose wedgefish *(Rhynchobatus laevis)*
- Scalloped hammerhead *(Sphyrna lewini)*
- Great hammerhead *(Sphyrna mokaran)*
- Hump-back mahseer *(Tor remadevii)*
- *Triplophysa kashmirensis*

5. Insects

- Pygmy hog-suckin' louse *(Haematopinus oliveri)*
- *Protostica myristicaensis*

6. Mammals

- ☆ Namdapha flying squirrel *(Biswamoyopterus biswasi)*
- ☆ Kashmir stag or hangul *(Cervus canadensis hanglu)*
- ☆ Elvira rat *(Cremnomys elvira)*
- ☆ Andaman shrew *(Crocidura andamanensis)*
- ☆ Jenkins' shrew *(Crocidura jenkinsi)*
- ☆ Nicobar shrew *(Crocidura nicobarica)*
- ☆ Sumatran rhinoceros *(Dicerorhinus sumatrensis)*
- ☆ Chinese pangolin *(Manis pentadactyla)*
- ☆ Leopard *(Panthera pardus)*
- ☆ Javan rhinoceros *(Rhinoceros sondaicus)*
- ☆ Malabar large-spotted civet *(Viverra civettina)* possibly extinct

7. Reptiles and Amphibians

- ☆ Dehradun stream frog *(Amolops chakrataensis)*
- ☆ Madras spotted skink *(Barkudia insularis)*
- ☆ Northern river terrapin *(Batagur baska)*
- ☆ Three-striped roofed turtle *(Batagur dhongoka)*
- ☆ Red-crowned roofed turtle *(Batagur kachuga)*
- ☆ Khasi hill rock toad *(Bufoides meghalaynus)*
- ☆ Adi's day gecko *(Cnemaspis adii)*
- ☆ *Cnemaspis anaikattiensis*
- ☆ Kottiyoor day gecko *(Cnemaspis kottiyoorensis)*
- ☆ *Cnemaspis shevaroyensis*
- ☆ Thackeray's dwarf gecko *(Cnemaspis thackerayi)*
- ☆ Chamba bent-toed gecko *(Cyrtodactylus chamba)*
- ☆ *Cyrtodactylus montanus*
- ☆ Hawksbill sea turtle *(Eretmochelys imbricata)*
- ☆ Ghats wart frog *(Fejervarya murthii)*
- ☆ Gharial *(Gavialis gangeticus)*
- ☆ Jeypore ground gecko *(Geckoella jeyporensis)*
- ☆ Satara gecko *(Hemidactylus sataraensis)*
- ☆ Southern Ghats slender gecko *(Hemiphyllodactylus aurantiacus)*
- ☆ *Hemiphyllodactylus kolliensis*
- ☆ Gundia Indian frog *(Indirana gundia)*
- ☆ Toad-skinned frog *(Indirana phrynoderma)*

- Yellow-headed tortoise *(Indotestudo elongata)*
- *Leptolalax khasiorum*
- Nagaland Asian toad *(Leptobrachella lateralis)*
- Asian forest tortoise *(Manouria emys)*
- Dzukou valley horned frog *(Megophrys dzukou)*
- Garo white-lipped horned frog *(Megophrys oreocrypta)*
- Rao's torrt frog *(Micrixalus kottigeharensis)*
- Spotted dancing frog *(Micrixalus specca)*
- Cave dancing frog *(Micrixalus spelunca)*
- Charles Darwin's frog *(Minervarya charlesdarwini)*
- Franky's narrow-mouthed frog *(Mysticellus franki)*
- *Nasikabatrachus bhupathi*
- Leith's softshell turtle *(Nilssonia leithii)*
- Black softshell turtle *(Nilssonia nigricans)*
- Dattatreya night frog *(Nyctibatrachus dattatreyaensis)*
- Indraneil's night frog *(Nyctibatrachus indraneili)*
- Assam roofed turtle *(Pangshura sylhetensis)*
- Cantor's giant softshell turtle *(Pelochelys cantorii)*
- Tura bubble-nest frog *(Philautus kempiae)*
- Kobo bubble-nest frog *(Philautus microdiscus)*
- Sacred grove bushfrog *(Philautus sanctisilvaticus)*
- Amboli bush frog (Pseudophilautus amboli)
- *Raorchestes aureus*
- White-spotted bush frog *(Raorchestes chalazodes)*
- Green eyed bushfrog *(Raorchestes chlorosomma)*
- Griet bush frog *(Raorchestes griet)*
- Kaikatti bushfrog *(Raorchestes kaikatti)*
- Mark's bushfrog *(Raorchestes marki)*
- Munnar bush frog *(Raorchestes munnarensis)*
- Ponmudi bush frog *(Raorchestes ponmudi)*
- *Raorchestes primarrumpfi*
- Resplendent shrubfrog *(Raorchestes resplendens)*
- Shillong bubble-nest frog *(Raorchestes shillongensis)*
- Anaimalai flying frog *(Rhacophorus pseudomalabaricus)*
- Sushil's bushfrog *(Raorchestes sushili)*
- *Rinophis goweri*

- ☆ Superb large fan-throated lizard *(Sarada superba)*
- ☆ Spiny lazy toad *(Scutiger spinosus)*
- ☆ Island pit viper *(Trimeresurus labialis)*
- ☆ Shevaroy hills earth snake *(Uropeltis shorttii)*
- ☆ Amboli toad *(Xanthophryne tigerina)*
- ☆ Ghats wart frog *(Zakerana murthii)*

II. Endangered

Listed by the IUCN as of the 2023-1 update.

1. Arachnids

- ☆ *Haploclastus kayi*
- ☆ *Poecilotheria formosa*
- ☆ *Poecilotheria miranda*
- ☆ *Poecilotheria rufilata*

2. Birds

- ☆ White-winged duck *(Asarcornis scutulata)*
- ☆ Forest owlet *(Athene blewitti)*
- ☆ Steppe eagle *(Aquila nipalensis)*
- ☆ Great knot *(Calidris tenuirostris)*
- ☆ Oriental stork *(Ciconia boyciana)*
- ☆ Saker falcon *(Falco cherrug)*
- ☆ Pallas's fish-eagle *(Haliaeetus leucoryphus)*
- ☆ Swamp grass babbler *(Laticlla cinerascens)*
- ☆ Greater adjutant *(Leptoptilos dubius)*
- ☆ Niligiri chilappan *(Montecincla cachinnans)*
- ☆ Banasura chilappan *(Montecincla jerdonil)*
- ☆ White-bellied blue robin *(Myiomela albiventris)*
- ☆ Nilgiri blue robin *(Myiomela major)*
- ☆ Egyptian vulture *(Neophron percnopterus)*
- ☆ White-headed duck *(Oxyura leucocephala)*
- ☆ Green peafowl *(Pavo muticus)*
- ☆ Manipur bush-quail *(Perdicula manipurensis)*
- ☆ Finn's weaver *(Ploceus megarhynchus)*
- ☆ Indian skimmer *(Rynchops albicollis)*
- ☆ Narcondam hornbill *(Rhyticero)*
- ☆ Nordmann's greenshank *(Tringa guttifer)*

- ☆ Black-chinned laughingthrush *(Trochalopteron jerdoni)*
- ☆ Nilgiri sholakili *(Sholicola major)*
- ☆ Great nicobar serpant-eagle *(Spilornis klossi)*
- ☆ Black-bellied tern *(Sterna acuticauda)*

3. Cnidaria

- ☆ *Acropora rudis*
- ☆ *Anacropora spinsosa*
- ☆ *Parasimplastrea sheppardi*
- ☆ *Porites desilveri*

4. Crustacean

- ☆ *Arachnochium kulsiense*
- ☆ *Macrobrachium lamarrei lamarroides*

5. Echinoderm

- ☆ *Holothuria lessoni*
- ☆ *Holothuria nobilis*
- ☆ *Holothuria scabra*
- ☆ Pineapple seacucumber *(Thelenota ananas)*

6. Fish (from 2018 - 135 as of 2023)

- ☆ Asian arowana *(Scleropages formosus)*
- ☆ Red line torpedo barb *(Sahyadria denisonii)*
- ☆ Golden mahaseer *(Tor putitora)*
- ☆ Deccan labeo *(Labeo potail)*

7. Insects

- ☆ Ludlow's Bhutan glory *(Bhutanitis ludlowi)*
- ☆ *Idionyx galeata*
- ☆ *Libellago balus*
- ☆ *Orthetrum andamanicum*

8. Mammals

- ☆ Red panda *(Ailurus fulgens)*
- ☆ Hog deer *(Axis porcinus)*
- ☆ Sei whale *(Balaenoptera borealis)*
- ☆ Blue whale *(Balaenoptera musculus)*

- Fin whale *(Balaenoptera physalus)*
- Banteng *(Bos javanicus)*
- Wild water buffalo *(Bubalus arnee)*
- Hispid hare *(Caprolagus hispidus)*
- Dhole *(Cuon alpinus)*
- Asian elephant *(Elephas maximus)*
- Indian elephant *(Elephas maximus indicus)*
- Onager *(Equus hemionus)*
- Woolly flying squirrel *(Eupetaurus cinereus)*
- Kelaart's long-clawed shrew *(Feroculus feroculus)*
- Manipur bush rat *(Hadromys humei)*
- Kolar leaf-nosed bat *(Hipposideros hypophyllus)*
- Nicobar leaf nosed bat *(Hipposideros nicobarulae)*
- Pomona roundleaf bat *(Hipposideros pomona)*
- Western hoolock gibbon *(Hoolock hoolock)*
- Mishmi hills hoolock gibbon *(Hoolock hoolock mishmiensis)*
- Salim Ali's fruit bat *(Latidens salimalii)*
- Eurasian lynx *(Lynx lynx)*
- White-cheeked macaque *(Macaca leucogenys)*
- Arunachal macaque *(Macaca munzala)*
- Lion-tailed macaque *(Macaca silenus)*
- Indian pangolin *(Manis crassicaudata)*
- Humpback whale *(Megaptera novaeangliae)*
- Kondana soft-furred rat *(Millardia kondana)*
- Kashmir musk deer *(Moschus cupreus)*
- Alpine musk deer *(Moschus chrysogaster)*
- Black musk deer *(Moschus fuscus)*
- White-bellied musk deer *(Moschus leucogaster)*
- Servant mouse *(Mus famulus)*
- Mandelli's mouse-eared bat *(Myotis sicarius)*
- Nilgiri tahr *(Nilgiritragus hylocrius)*
- Bengal slow loris *(Nycticebus bengalensis)*
- Irrawaddy dolphin *(Orcaella brevirostris)*
- Asiatic lion *(Panthera leo persica)*
- Bengal tiger *(Panthera tigris tigris)*
- Ganges river dolphin *(Platanista gangetica)*

- Indus river dolphin *(Platanista minor)*
- Pygmy hog *(Porcula salvania)*
- Nicobar flying fox *(Pteropus faunulus)*
- Nonsense rat *(Rattus burrus)*
- Kerala rat *(Rattus ranjiniae)*
- Andaman horseshoe bat *(Rhinolophus cognatus)*
- Sangai *(Rucervus eldii)*
- Kashmir gray langur *(Semnopithecus ajax)*
- Indian Ocean humpback dolphin *(Sousa plumbea)*
- Day's shrew *(Suncus dayi)*
- Gee's golden langur *(Trachypithecus geei)*
- Phayre's leaf-monkey *(Trachypithecus phayrei)*
- Trachypithecus phayrei phayrei
- Capped langur *(Trachypithecus pileatus brahma)*
- Tenebrous capped langur *(Trachypithecus pileatus tenebricus)*
- Blond-bellied langur *(Trachypithecus pileatus pileatus)*
- Nicobar treeshrew *(Tupaia nicobarica)*
- Nilgiri long-tailed tree mouse *(Vandeleuria nilagirica)*

9. Mollusk

- *Cremnoconchus carinatus*
- *Cremnoconchus syhadrensis*
- *Pseudomulleria dalyi*

10. Reptiles and Amphibians (from 2018 - 122 as of 2023)

- Perrotet's vine snake (*Ahaetulla perroteti*)
- Green turtle *(Chelonia mydas*)
- Indian narrow-headed softshell turtle *(Chitra indica*)
- Goan day gecko *(Cnemaspis goaensis*)
- Wyanad day gecko *(Cnemaspis wynadensis*)
- Keeled box turtle *(Cuora mouhotii*)
- Boulenger's dasia *(Dasia subcaerulea*)
- Poona skink *(Eurylepis poonaensis*)
- Inger's mabuya *(Eutropis clivicola*)
- Indian kangaroo lizard *(Otocryptis beddomii*)
- Travancore Hills thorntail snake *(Platyplectrurus madurensis*)
- Travancore earth snake *(Rhinophis travancoricus*)

- ☆ Cochin forest cane turtle *(Vijayachelys silvatica*)

III. Vulnerable

Listed by the IUCN As of 2012. As of the 2023-1 update, there are 481 animals listed as vulnerable.

1. Birds

- ☆ Sarus crane (*Antigone*)
- ☆ Nicobar megapode (*Megapodius nicobariensis*)
- ☆ Dalmatian pelican (*Pelecanus crispus*)

2. Mammals

- ☆ Cheetah (*Acinonyx jubatus*)
- ☆ Brown bear (*Ursus arctos*)
- ☆ Gaur (*Bos gaurus*)
- ☆ Four-horned antelope or chousingha (*Tetracerus quadricornis*)
- ☆ Oriental small-clawed otter (*Aonyx cinerea*)
- ☆ White-chested bear (*Ursus thibetanus*)
- ☆ Yak (*Bos grunhniens*)
- ☆ Takin (*Budorcas taxicolor*)
- ☆ Barasingha (*Rucervus duvaucelii*)
- ☆ Clouded leopard (*Neofelis nebulosa*)
- ☆ Dugong (*Dugong dugon*)
- ☆ Sun bear (*Helarctos malayanus*)
- ☆ Stump-tailed macaque (*Macaca arctoides*)
- ☆ Marbled cat (*Pardofelis marmorata*)
- ☆ Sperm whale (*Physeter macrocephalus*)
- ☆ Rusty-spotted cat (*Prionailurus rubiginosus*)
- ☆ Indian rhinoceros (*Rhinoceros unicornis)*
- ☆ Snow leopard (*Uncia uncia)*
- ☆ Nilgiri marten (*Martes gwatkinsii*)
- ☆ Sloth bear (*Melursus ursinus*)

3. Reptiles and Amphibians

- ☆ Olive ridley sea turtle (*Lepidochelys olivacea*)

6. Threats to Biodiversity

The loss of biodiversity is a significant concern. Species are becoming extinct at the fastest rate known in geological history and most of these extinctions have been tight to human activity. The core threat to biodiversity on the planet and therefore a threat to human welfare, is the combination of human population growth and resource exploitation. The highest proximate threats to biodiversity are habitat loss and destruction, agricultural conservation, urbanization, alterations in ecosystem composition, over-exploitation, pollution, climate change and overharvesting.

i) Habitat Loss and Destruction

Habitat destruction or habitat loss is the process by which a natural habitat becomes incapable of supporting its native species. The organisms that previously inhabited the site are displaced or dead, thereby reducing biodiversity and species abundance. The primary effect of habitat destruction is a reduction in biodiversity, which refers to the variety and abundance of different species of animals and plants in a particular setting. When an animal loses the natural home or habitat that it needs to survive, its numbers decline rapidly, and it moves toward extinction. Usually as a direct result of human activity and population growth, is a major force in the loss of species, populations and ecosystems.

ii) Agricultural Conversion

The *change* is motivated by the fact that *loss* of habitat due to *agricultural conversion* is argued to be the main driver of *biodiversity loss.* Agricultural activities such as tillage, drainage, intercropping, rotation and extensive usage of pesticides and fertilizers have significant effects for wild species of flora and fauna.

iii) Urbanization

Urbanization has an important impact on biodiversity, mostly driving changes in species assemblages, through the replacement of specialist with generalist species, thus leading to biotic homogenization. Mobility is also assumed to greatly affect species' ability to cope in urban environments. It leads to habitat fragmentation, where larger continuous habitats are divided into smaller unconnected patches. It also causes habitat loss, through an increase in roads and buildings that are not producing any biomass.

iv) Alterations in Ecosystem Composition

Alterations in ecosystem composition such as the loss or decline of a species, can lead to a loss of biodiversity. The introduction of exotic (non-native) species can disrupt entire ecosystems and impact populations of native plants or animals. These invaders can adversely affect native species by eating them, infecting them, competing with them, or mating with them. For example, efforts to eliminate coyotes in the canyons of southern California are linked to decreases in song bird populations in the area. As coyote populations were reduced, the populations of their prey, primarily raccoons, increased. Since raccoons eat bird eggs, fewer coyotes led to more raccoons eating more eggs, resulting in fewer song birds.

v) Over-Exploitation

Overexploitation means harvesting species from the wild at rates faster than natural populations can recover. Natural resource consumption rates and human population size exert tremendous pressure on the world's plants and animals. Unmanageable hunting, fishing, logging, or gathering of wild populations leads to their commercial, ecological, or global extinction. Most of the land animals we eat are farm-raised, not hunted in the wild. Fish, on the other hand, are often still fished from the wild.

vi) Pollution

All forms of pollution pose a serious threat to biodiversity. Human-generated pollution and contamination can affect all levels of biodiversity. Air pollution affects all things. It is harmful to our health, and it impacts the environment reducing visibility and blocking sunlight, causing acid rain and harming forests, wildlife and agriculture. Greenhouse gas pollution, the cause of climate change, affects the entire planet. Eutrophication, the process of accumulation of nutrients, including nitrogen, in water bodies, often results from air pollution. Nutrient overloads in aquatic ecosystems can cause algae blooms and ultimately a loss of oxygen, and it leads to absence of life.

vii) Climate Change

Global climate change can alter environmental conditions. Species and populations may be lost if they are unable to adapt to new conditions or relocate. The environmental changes being driven by climate change are disturbing natural habitats and species in ways that are still only becoming clear. There are signs that rising temperatures are affecting biodiversity, while changing rainfall patterns, extreme weather events, and ocean acidification are putting pressure on species already threatened by other human activities.

viii) Overharvesting

Overharvesting or overfishing in the case of fish and marine invertebrates, depletes some species to very low numbers and drives others to extinction. In practical terms, it reduces valuable living resources to such low levels that their exploitation is no longer sustainable. Overharvesting threatens biodiversity by degrading ecosystems and reducing species of plants, animals and other organisms. The unmanageable use of natural resources and overexploitation, which occurs when harvesting exceeds reproduction of wild plant and animal species, continues to be a major threat to biodiversity.

7. Biodiversity Conservation

Biodiversity conservation is the protection and managing of biodiversity to obtain resources for sustainable development. Biodiversity conservation has to preserve the diversity of species, sustainable utilization of species and ecosystem, and to maintain life-supporting systems and essential ecological processes. Biodiversity refers to the variability of life on earth. It can be conserved in the two ways:

1. In-situ Conservation

In-situ conservation of biodiversity is the conservation of species within their natural habitat. In this method, the natural ecosystem is maintained and protected. The in-situ conservation has several advantages. Certain protected areas where in-situ conservation takes place include national parks, wildlife sanctuaries and biosphere reserves. Following are the advantages of in-situ conservation:

- It is a cost-effective and convenient method of conserving biodiversity.
- A large number of living organisms can be conserved simultaneously.
- Since the organisms are in a natural ecosystem, they can evolve better and can easily adjust to different environmental conditions.

i) National Parks

These are small reserves maintained by the government. Its boundaries are well demarcated and human activities such as grazing, forestry, habitat and cultivation are prohibited. For *e.g.,* Kanha National Park, Bandipur National Park.

ii) Wildlife Sanctuaries

These are the regions where only wild animals are found. Human activities such as timber harvesting, cultivation, collection of woods and other forest products are allowed here as long as they do not interfere with the conservation project. Also, tourists visit these places for recreation.

iii) Biosphere Reserves

Biosphere reserves are multi-purpose protected areas where the wildlife, traditional lifestyle of the inhabitants and domesticated plants and animals are protected. Tourist and research activities are permitted here.

2. Ex-situ Conservation

Ex-situ conservation of biodiversity involves the breeding and maintenance of endangered species in artificial ecosystems such as zoos, nurseries, botanical gardens, gene banks, *etc.* There is less competition for food, water and space among the organisms. Ex-situ conservation has the following advantages:

- The animals are provided with a longer time and breeding activity.
- The species bred in captivity can be reintroduced in the wild.
- Genetic techniques can be used for the preservation of endangered species.

8. Strategies for Biodiversity Conservation

Following are the important strategies for biodiversity conservation:

- ☆ Develop and maintain protected areas like national park, sanctuaries to protect the endangered and critical species.
- ☆ All the vulnerable animals should be identified and conserved.
- ☆ Efficient utilization of the resources.
- ☆ Focus on reforestation, wetland restoration, and other habitat rehabilitation efforts.
- ☆ Hunting and Poaching of wild animals should be prevented.
- ☆ Sustainable agriculture practice should be followed.
- ☆ Local community can be involved in the conservation efforts.
- ☆ Raise knowledge about the importance of biodiversity conservation.

9. Conclusion

Biodiversity help in maintaing stability and resilience of the ecosystem. Ecosystem also provides various services like air, water, and food. Therfore, our survival is dependent on it. Many medicines are derived from plants and animals. Biodiversity act as a reservoir of genetic resource. It provides various services like pollination, soil fertility, and climate regulation. The richness of the species, the ecosystem, the environment and the sustainable growth of life on earth is important. Biodiversity conservation is essential to reduce the activities leading to pollution so as to conserve biodiversity. Due to deforestation, there is the loss of habitat. Due to this reason, wild animals are unable to survive in the new environment and die. India's 12 national biodiversity targets include creating awareness about biodiversity, enforcing policies to document, research and conserving biological resources.

10. References

1. Archived from the original on 14 April 2012. Retrieved 22 June 2012.
2. Darwin C R. On the origin of species by means of natural selection. London: John Murray. 1859.
3. Matthew Knight (19 June 2012). "Extinction threat 'a call to world leaders' at Rio Earth Summit". CNN. Retrieved 24 January 2024.
4. https: //www.montana.edu/president/medal/medal-eowilson.html
5. https: //upsccolorfullnotes.com/critically-endangered-species-in-india-2021/
6. https: //www.hindustantimes.com/india-news/number-of-indian-species-in-endangered-list-going-up/story-qPqyu8T4R09xFCFA7a6GAL.html
7. https: //www.worldwildlife.org/species/directory?direction=desc and sort=extinction_status

8. https: //www.zmescience.com/ecology/resources-climate-59447/
9. https: //www.nrcs.usda.gov/Internet/FSE_DOCUMENTS/nrcs142p2_050947.pdf
10. https: //www.who.int/news-room/fact-sheets/detail/biodiversity-and-health
11. https: //www.ncbi.nlm.nih.gov/pmc/articles/PMC6871202/
12. https: //byjus.com/biology/biodiversity-conservation/
13. https: //www.ecowatch.com/why-biodiversity-matters-2646065739.html#: ~: text=Biodiversity per cent 20Helps per cent 20Fight per cent 20Disease. and text=This per cent 20means per cent 20that per cent 20every per cent 20time,as per cent 20Lyme per cent 20disease per cent 20and per cent 20malaria.
14. https: //www.weforum.org/agenda/2020/05/5-reasons-why-biodiversity-matters-human-health-economies-business-wellbeing-coronavirus-covid19-animals-nature-ecosystems/
15. https: //byjus.com/questions/why-is-india-called-a-mega-diversity-centre/
16. https: //byjus.com/biology/biodiversity-conservation/
17. https: //www.in.undp.org/content/india/en/home/climate-and-disaster-reslience/successstories/IBA2018.html
18. https: //www.toppr.com/guides/essays/essay-on-biodiversity/
19. https: //www.ecowatch.com/why-biodiversity-matters-2646065739.html
20. https: //www.pmfias.com/biodiversity-hotspots-india/
21. https: //www.britannica.com/place/western-Himalayas
22. https: //www.globalissues.org/article/170/why-is-biodiversity-important-who-cares
23. https: //www.iucnredlist.org. Accessed on 23 January 2024.
24. Red list has 132 species of plants, animals from India
25. The IUCN Red List of Threatened Species. Version 2023.
26. Vidhya V. 2022. Megabiodiversity of India: Endangered species, Threats and its Conservation. Life Sciences: Trends and Technology. (56-69).

Chapter 3

The Imperative of Biodiversity Conservation: Understanding the Complex Interplay of Genetic, Species, and Ecosystem Diversity

V. Sundara Prabha

Associate Professor, Department of Botany, S.T. Hindu College Nagercoil, Affiliated to MS University, Tirunelveli– 629002, Tamil Nadu
e-mail: prabhavasu90@gmail.com

ABSTRACT

Biodiversity refers to the variety of life on Earth and is important for human well-being due to the services provided by them on which entire nations depend. Unfortunately, human actions such as deforestation, degradation, global warming, pollution, and caps on resources endanger biodiversity, leading to the loss of animal species and extinction. Restoring and safeguarding biodiversity calls for an elaborate comprehension of its genetic, species, and ecosystem interrelations. This calls for the need to protect biodiversity and its many levels, understand the key issues that can harm it, and formulate policies for its sustainable use to balance and maintain humanity's essential services.

Keywords: *Biodiversity, Pollution, Global warming and Extinction*

1. Introduction

Biodiversity is derived from two Latin words-bios means life and diversitas means variety. Biodiversity is thus defined as the 'Full variety of Life on Earth', in other words, it refers to the totality of different kinds of living species, their forms, levels and combinations on earth. The term biodiversity was coined by W.G. Rosen (1985). Biological diversity may be defined as the variation that is present in all species of plants and animals, their genetic material and the ecosystems in which they live. Diversity is of three levels. They are genetic diversity where the variations are seen in genes and their genotypes, species diversity which includes species richness and ecosystem diversity with all communities of species along with their environment. The importance of biodiversity for humankind has been well recognized in recent decades and many would argue that diversity is essential for allowing sustainable development of various human activities. Biological diversity can enable social and economic systems to flourish in ways that allow the poorest to meet their food and nutritional needs and retain the cultural diversity of countries throughout the world (Shiva, 1994). The simplest definition of biodiversity is the variation of life at all levels of biological organ station (Antonelli *et al.,* 2018). The term biodiversity generally includes Genetic diversity, Species and Ecosystem diversity.

The diverse nature of life on earth has provided for human needs over thousands of years. This diversity of living organisms forms a supporting system that has been used for civilization for its growth and development. The people who used this "bounty of nature" usefully, and carefully are sustainably survived. Those who overused and misused diversity make it disintegrate.

India is one of the world's most biodiverse countries in the world with a wide variety of species. India's huge diverse climatic conditions and geographic nature support many ecosystems, including forests, wetlands, grasslands, deserts, and marine environments. India is a diverse country which is situated north of the equator between 66°E to 98°E long. and 8°N to 36°N lat. It is bordered by Nepal, China and Bhutan in the north; Bangladesh and Myanmar in the east; the Bay of Bengal in the southeast; the Indian Ocean in the south; the Arabian Sea in the west; and Pakistan in the northwest. The total land area of India is 3.3 million square kilometers and it is the seventh largest country in the world. It is 2,933 kms wide and 3,214 kms long. India is one of 17 mega-biodiverse countries in the world. With only 2.4 percent of the earth's land area, it accounts for 7-8 percent of the world's recorded species. Home to 96,000 species of animals, 47,000 species of plants and nearly half the world's aquatic plants, India's management of its natural resources is crucial to protecting global biodiversity.

2. Types of Biodiversity

2.1. Genetic Diversity

Every member of a species differs widely from other individuals in its genetic makeup because of the large number of combinations possible in the genes that give every individual specific characteristics. Thus, for example, each plant is very different from all others. This genetic variability is essential for a healthy breeding population of a species. If the number of breeding individuals is reduced, the dissimilarity of genetic makeup is reduced and in-breeding occurs. Eventually, this can lead to the extinction of the species. The diversity in wild species forms the 'gene pool' from which our crops and domestic animals have been developed over thousands of years. Today the variety of nature's bounty is being further harnessed by using wild relatives of crop plants to create new varieties of more productive crops and to breed better domestic animals. Using modern biotechnology manipulates genes to develop better types of medicines and a variety of industrial products.

The foundation of survival and adaptation is genetic variety, which enables the continuation and advancement of the adaptive processes necessary for evolutionary success and also for human existence. Time, space, and fitness can all be used to analyze the survival and adaptation of living organisms. Adaptability, genetic stability, and variability are all components of fitness. Extinction may be brought on by biotic or abiotic stressors, such as competition, predation, parasitism, and illness, or by habitat alteration and isolation brought on by natural disasters, human activity, or gradual geological and climatic change. The genetic variety of plant genetic resources must be adequately recognized, maintained, and utilized in light of these risks.

2.2. Species Diversity

The species diversity of a region is the total number of plant and animal species found there. This diversity can be observed in both agricultural and natural ecosystems. There are species in richer areas than others. Compared to plantations created by the Forest Department for the production of timber natural undisturbed tropical forests have a far higher species diversity. The ecosystem of a natural forest provides. Numerous non-wood products including fruit fuel wood fodder fiber gum resin and medications are essential to the local population wood. The wide range of products that are necessary for local consumption are not available on plantations. The economic benefits of non-wood forest products are thought to outweigh the costs of clearing a forest for lumber in the long run. Thus a natural forest value has a far higher species diversity than a plantation. Crop diversity in contemporary intensive agricultural ecosystems is comparatively lower compared to conventional agropastoral farming systems which planted a variety of crops.

2.3. Ecosystem Diversity

There are many distinct ecosystems on the planet each of which has its complement. Different species are linked together according to habitat differences. Diverse ecosystems can be explained for a particular geographic area or a political body like a nation a state or a taluka. Landscapes like forests grasslands and deserts are examples of distinctive ecosystems mountains *etc.* in addition to aquatic ecosystems such as the sea lakes and rivers each area has areas that have been altered by humans like farmland or pastures used for grazing. When it occurs an ecosystem is called natural is largely untouched by human activity or odified refers to the substitution of other forms uses including urban areas and farms. Wilderness areas have the most natural ecosystems. If natural ecosystems are overused or misused their productivity eventually decreases and they are then said to be degraded. India is exceptionally rich in its ecosystem diversity.

3. Ecosystem Diversity has Three Outlooks

3.1. Alpha (α) Diversity

It is the biodiversity within a particular region, community or ecosystem. It is usually expressed by the number of species in that ecosystem. This can be calculated by including the number of distinct groups of organisms (taxa) within the ecosystem (Families, genera and species).

3.2. Beta (β) Diversity

Beta diversity is a measure of biodiversity that works by matching up the species diversity between ecosystems or along environmental pitches. This involves matching the number of taxa that are exclusive to each of the ecosystems. It depicts a range of communities due to the substitution of species which takes place due to the occurrence of diverse microhabitats, niches and environmental conditions.

3.3. Gamma (γ) Diversity

It illustrates the diversity of habitat over a total geographical area. It is a product of component ecosystems (alpha diversity) and the between component ecosystems (beta diversity).

4. The Distribution of Biodiversity by Geography

The world's biodiversity is not dispersed equally. Tropical areas especially those close to the equator are typically the richest because of their warm temperatures and high primary productivity which support a diverse range of species (Gaston 2000). Because of the favorable conditions for a variety of marine life marine biodiversity is also particularly high along the Western Pacific coasts and in mid-latitudinal bands across all oceans.

4.1. Uneven Biodiversity Distribution Across the Earth

Due to the tilt of its axis relative to the sun and the curvature of the Earth, different areas of the planet vary in the amount of sunlight energy received throughout the year. Because of this the duration of different seasons (warm, cold, wet, and dry) in different areas, as well as other environmental factors (temperature, humidity *etc.*) varies. Due to this variation, the distribution of biodiversity varies establishing different hotspots.

One of the most important outcomes of different affecting factors in different regions on the planet has created particular sets of environmental settings, which results in variation in predominant vegetation. The species living in different regions are distinguished by the adaptations that allow survival under the particular set of environmental conditions of the region and establish biomes. A biome is a biological population of organisms related to particular climactic and geographic conditions. These regions can be broadly classified into terrestrial biomes and aquatic ecosystems in which the biodiversity distribution varies. The important factors that affect the distribution of biodiversity. They are Climate changes, Topography of the earth, Soil structure, Type of other species present in the world, geological age, habitat size of the habitat, Latitude and Taxon

4.2. Biodiversity Trends and Hotspots

Biodiversity hotspots are geographic areas with an exceptionally high richness of species, including rare and endemic species. Such hotspots have deep evolutionary roots and are concentrated in areas where past climatic variability was moderate (Enquist *et al.,* 2019; Brown *et al.,* 2020; Trew and Maclean, 2021). Many studies have proposed biodiversity hotspots based on different criteria, taxa and geographic contexts.

According to current trends, habitat loss climate change and other anthropogenic pressures may cause this increase to slow down in the future (Robosky 2009). The term 'biodiversity hotspot' was coined by Norman Myers in the late 1980s who identified 18 geographical regions as conservation priorities because they contained large numbers of endemic species found in relatively small areas that were facing significant threats of habitat loss. It was reasoned that protecting hotspots, defined in this manner, should prevent the extinction of a larger number.

The study of the distribution of species, which has long been a central focus of ecology and biogeography, is taking on new urgency as evidence of the global biodiversity crisis mounts. The question of what geographical regions protect to maintain the most biological diversity is central to the design of effective conservation programs

4.3. Convention on Biological Diversity

The Convention on Biological Diversity's founding. An important turning point in global environmental governance was the 1972 Stockholm UN Conference on the Human Environment which established the framework for international collaboration on biodiversity and conservation issues. This conference sought to address the growing concerns about environmental degradation and the need for coordinated action and one of its main outcomes was the creation of the UN Environment Programme (UNEP)

The Convention on Biological Diversity (CBD) was negotiated and signed by nations at the Earth Summit in Rio de Janeiro, Brazil on June 5, 1992. The convention came into force on December 29, 1993. India became a party to the convention on February 18, 1994. At present, there are 196 Parties to this Convention.

The three main objectives of this convention are the conservation of biological diversity, sustainable use of natural resources and fair and equitable sharing of benefits arising out of the use of these resources and associated traditional knowledge.

UN Convention on Biological Diversity (CBD) was adopted in 2010 for the following Vision for its Strategic Plan. By 2050, biodiversity will be valued, conserved, restored and wisely used, maintaining ecosystem services, sustaining a healthy planet and delivering benefits essential for all people. The mission of CBD is to take effective and urgent action to halt the loss of biodiversity to ensure that by 2020 ecosystems are resilient and continue to provide essential services, thereby securing the planet's variety of life, and contributing to human well-being, and poverty eradication (Decision CBD/COP/X/2).

4.4. Advantages of Biodiversity

Biodiversity or the diversity of life on Earth offers many advantages that are vital to ecosystem health and human societies. The following are some of the main advantages of biodiversity.

Biodiversity contributes to a wide range of ecosystem services that are vital for human survival and quality of life in the ecosystem service including Fuel fiber, freshwater food and genetic resources are examples of provisioning services. Various crops livestock and fish are produced by diverse ecosystems and are vital for livelihoods and nutrition. ICUN (2012).

Controlling Services along with biodiversity aids in the control of disease the climate and the quality of the air and water. Wetlands filter pollutants forests store carbon and a variety of plant species can improve the stability and fertility of the soil.

Biodiversity provides cultural Services which enhance recreational opportunities spiritual values and cultural identity. Wildlife and natural settings support tourism education and cultural heritage. The underlying mechanisms that sustain ecosystems such as pollination soil formation and nutrient cycling are known as supporting services. For instance, bees butterflies and other pollinators are crucial to the reproduction of numerous plants including crops.

Stability and Resilience of biodiverse ecosystems tend to be more resilient to shocks such as human activity climate change and natural disasters. Diverse species that can adapt to shifting conditions and recover from disturbances result in more stable ecosystems over time.

Financial Gains among the commercial activities that biodiversity promotes are forestry fishing tourism and agriculture. The sustainability and productivity of these industries can be increased by healthy ecosystems which will increase their economic opportunities and generate employment. Ecotourism for example depends on the preservation of wildlife and natural environments.

A lot of medications are made from natural substances that can be found in microbe plants and animals. Genetic material for pharmaceutical development comes from biodiversity and the extinction of species may hinder the development of new medications. For many communities, traditional medical knowledge about medicinal plants is also very important.

Food Security offers a range of crops and livestock that can adjust to shifting environmental conditions biodiversity helps ensure food security, genetic variety among species (*e.g.* 3. Breeding programs that aim to increase resilience to pest's diseases and climate change require a variety of crops. Because they absorb carbon dioxide and affect regional weather patterns biodiverse ecosystems–like forests and wetlands–have a major impact on climate regulation. To lessen the effects of climate, change these ecosystems must be protected. Diverse plant species improve soil structure inhibit erosion and improve nutrient cycling all of which are factors in soil health. A healthy soil is necessary for both agriculture and the preservation of ecosystem services.

Aesthetic and Cultural Value: Biodiversity confers aesthetic pleasure and enhances human experiences. Biodiversity wildlife and natural landscapes all support mental health leisure and artistic and cultural inspiration. Many natural and anthropogenic (caused by humans) factors are seriously endangering biodiversity. Estimates indicate that the current extinction rates are almost 1000 times higher than the natural background rate and if current trends continue, they could rise to 10000 times that rate in the next century. The devastating extinctions of the past might be mirrored in this possible loss of biodiversity.

4.5. Major Threats to Biological Diversity

Habitat destruction is the most important threat to biological diversity is habitat destruction and fragmentation due to urbanization, agriculture, deforestation and infrastructure development. When natural habitats are converted for human use, these ecosystem-dependent types often fail to survive.

Accelerated climate change is a serious threat to biological diversity, with sedimentation patterns that can alter habitats, disrupt migration patterns, and lead to the extinction of species that cannot be adapted quickly enough There is. Assessment of the Millennium Ecosystem and various reports from IUCN and UNEP highlight the impact of climate change on biological diversity.

Contamination by agricultural wastewater, industrial waste and plastic waste can affect ecosystems and damage species. Chemicals such as pesticides and heavy metals can accumulate in food chains, providing toxic effects on wild animals and humans.

Uncontinuous harvesting of resources, including overfishing, logging and poaching can lead to population decline and species extinction. The demand for wildlife products such as ivory and traditional drugs exacerbate this threat.

The introduction of non-local species can interfere with local ecosystems, transcend local species, and lead to reduced biodiversity. Invasive species can change the structure and function of their habitat, making it difficult for local species to thrive.

The introduction of genetically modified organisms (GMOs) and plants can represent risks to native species and ecosystems. These organisms can transcend local species or introduce new genes that disrupt local genetic varieties.

Geological events such as earthquakes, volcanic eruptions, and storm events (hurricanes, floods, *etc.*) can have immediate, catastrophic effects on biological diversity. These events are natural, but frequency and intensity can be tightened by climate change.

4.6. Endangered and Endemic Plant Species of India

The endemic and endangered species of India are important to understand the wide variety of plant species that are found in the country. Of these species, there are several of which are endangered by human activity. The endangered species in the country are categorized as Vulnerable, Rare, Indeterminate and Threatened species based on their level of extinction threat, usually using a system like the IUCN Red List. Many plant species are now increasingly threatened due to changes in their habitats induced by human activity. Apart from major trees, shrubs and climbers that are extremely habitat-specific and thus endangered, there are thousands of small herbs which are greatly threatened by habitat loss. Several orchids are yet another group of plants that are under threat.

The flora of India is one of the richest in the world due to the country's wide range of climate, topology, and environment. There are over 15,000 species of flowering plants in India which account for 6 per cent of all plant species in the world. Many plant species are being destroyed, however, due to their prevalent removal. Roughly 1/4 of all plant species in the world are at risk of being endangered or going extinct. The combination of global warming and habitat destruction is the sole reason for the disappearance of many plants. Though there are thousands of interesting and unusual plants, here are some common plants that have become rare and endangered species in the past 30 years due to habitat destruction. In India, 450 plant species have been identified as endangered species. India's biodiversity is threatened primarily due to Habitat destruction, Degradation and over-exploitation of resources. *Abutilon indicum, Acacia planifrons, Actinodaphne lawsonii, Ceropegia odorata, Chlorophytum malabaricum, Diospyros celibica, Pterospermum reticulatum etc.* are some of the endangered plant species. The most important factors affecting endangered species is that Human beings dispose of wastes indiscriminately in nature thereby polluting the air, land and water. These pollutants enter the food chain and accumulate in living creatures resulting in death. Natural resources are Over-exploitation which leads to their extinction. Climate change is brought about by the accumulation of greenhouse gases in the atmosphere. Climate change threatens organisms and ecosystems and they cannot adjust to the changing environmental conditions leading to their death and extinction.

5. Biodiversity Conservation

Conservation is the management of the biosphere for the benefit of all living things including humans to preserve its capacity to support the needs and ambitions of future generations while producing long-term benefits for the current generation. In detail, conservation is the process of maximizing benefits from the biosphere while preventing it from degradation. Every species in the natural world is cooperative either directly or indirectly in addition to being supportive of one another. Therefore, the resource that family community countries and future generations rely on is biodiversity. All living things on Earth are connected by forming a separate ecosystem where all biological species interact with one another. It is the web of life to put it another way. Both in-situ and ex-situ conservation can contribute to this.

In situ conservation refers to the preservation of species in their native environments within protected areas such as national parks sanctuaries and biosphere reserves. In ex situ conservation endangered species are maintained and bred in botanical gardens seed banks gene banks and other partially or completely controlled environments. Among other things it covers tissue culture and cryopreservation. The ability of a given habitat to support life is indicated by its rich biodiversity. Biodiversity-rich areas are designated as national parks wildlife sanctuaries biosphere reserves and ecologically sensitive and fragile areas to conserve and protect biodiversity. Additional tactics include reducing the strain on re-served forests through alternative fuel wood and fodder measures reforesting degraded areas and wetlands and utilizing specific ex-situ conservation facilities like gene banks.

Articles 48a and 51a(g) of the Indian Constitution also protect biodiversity. Several laws about biodiversity were enforced by the federal and state governments under this article including the Forest Act of 1927 the Wildlife (Protection) Act of 1972 the Forest (Conservation) Act of 1980 and the Environment (Protection) Act of 1986. Numerous state laws and statutes of forests and other natural resources support the various central Acts. Governments and NGOs will start new ambitious projects for conservation and protection. Several scientists worldwide have suggested the following crucial actions to preserve current wildlife:

- Conservation and preservation of existing natural forests and aquatic ecosystems (Agarawal 2009 and Kotwal and Banerjee 1998).
- Particular care should be taken to protect species that are rare endangered or vulnerable.
- To protect wildlife in their natural habitats or in artificial habitats like zoos and botanical gardens proper planning of land and water use is necessary.

- Wildlife products of wild plants and animals used for international trade should only be permitted at levels that would not endanger the related wildlife.
- Conservation efforts should be made to preserve as many types of food crops as possible including forage plants timber trees livestock aquaculture animals and domesticated animals.
- Disease control especially about wild plants and animals. To prevent poaching illegal hunting and the trade in animal products strict laws must be enacted and upheld.
- The preservation and enhancement of extensive ecosystems.

5.1. The Importance of Biodiversity Conservation

Biodiversity is essential for maintaining ecosystem stability. Every species in an ecosystem contributes and helps the system to function as a whole. A species extinction can upset the equilibrium and have detrimental effects like decreased productivity weakened resilience and heightened susceptibility to illnesses. An ecosystem's capacity to react and adjust to changes in their environment is facilitated by biodiversity. The diversity of species enhances the possibility that some will be able to survive and thrive in changing environments, ensuring the overall resilience of the ecosystem.

Human well-being depends on many ecological services provided by biodiversity. Pollination, soil creation, water purification, nutrient cycling, and climate regulation are some of these services. Decreased biodiversity may be detrimental. The diversity of crops and animal genetic resources provided by biodiversity can be used to feed and nourish the growing human population. Agrobiodiversity is also crucial for enhancing climate change adaptation and agricultural resilience productivity. Traditions and ideas about natural resources demonstrate the close relationship between a society's biological and cultural variety. Identity preservation and community cohesion are facilitated by biodiversity's cultural and spiritual value.

Biodiversity is a significant source of income and a means of subsistence for millions of people. Wildlife tourism, traditional medicine, and the selling of non-timber forest products provide employment and income-generating options for many rural people. The diversity of species and ecosystems is the foundation for resilience and climate change adaptation. Due to unique genetic traits, some species are more resistant to environmental stresses like heat and drought.

6. Conclusion

Biodiversity is essential to ecosystem health, human well-being, and economic stability. Ecosystem conservation is essential to maintaining livelihoods, maintaining ecological balance, and ensuring ecosystems' resilience to environmental changes. Effective conservation methods, legal protections, and public awareness campaigns are necessary to preserve the planet's biodiversity for future generations. There is no one primary driver of biodiversity loss or stability because it is a very broad, intricate, and interdependent process. For the benefit of humanity, biological variety must be properly documented to be conserved and used sustainably. Additionally, satellite remote sensing monitoring of the changes is required to determine the indicators of biodiversity changes.

7. References

1. https: //www.nature.com/articles/s41467-021-24207-4
2. Biodiversity: Concept, Threats and Conservation, December 2015, Environment Conservation Journal 16(3): 19-28.
3. Biological Diversity: Protecting the variety of life on Earth Elsa Tsioumani, Ph.D. Asterios Tsioumanis, Ph.D. September 2020
4. https: //www.earthday.org/the-importance-of-mangrove-forests/
5. https: //www.iisd.org/system/files/2020-09/still-one-earth-biodiversity.pdf
6. https: //www.researchgate.net/publication/317313069_Biodiversity-Threats_and_Conservation
7. https: //ebooks.inflibnet.ac.in/esp12/chapter/current-status-of-environment-in-india-biodiversity-and-climate-change-iii/
8. India's Fifth National Report to The Convention on Biological Diversity, MoEF, GOI
9. Ministry of Environment and Forests and Climate Change, Government of India http: //www.moef.nic.in/division/ramsar-convention-wetland
10. India State of Forest Report (ISFR) 2015-16
11. National Wildlife Database Cell, Wildlife Institute of India
12. Gaston KJ. Global patterns in biodiversity. Nature. 2000 May 11;405(6783): 220-7
13. Myers, N., Mittermeier, R., Mittermeier, C. *et al.,* Biodiversity hotspots for conservation priorities. *Nature* **403**, 853–858 (2000)
14. McPeek Mark A., Resetarits William J. and Holt Robert D. 2024The evolution of passive dispersal versus habitat selection have differing emergent consequences in metacommunitiesPhil. Trans. R. Soc. B37920230126
15. Daniel L. Rabosky. Ecological limits and diversification rate: alternative paradigms to explain the variation in species richness among clades and regions, Ecology Letter. Volume12, Issue8, August 2009, Pages 735-743.
16. Arjjumend, Hasrat and Koutouki, Konstantia and Siddiqui, Shoaib Alam, Evolution of International Governance of Biodiversity (2016). Journal of Global Resources Volume 3 July 2016 Page 1-15

17. IUCN Red List categories and criteria, version 3.1, second edition, 2012
18. Nick Salafsky, Claire Relton, Bruce E. Young, Philippe Lamarre, Monika Bohm, Maxime Chénier, Erica Cochrane, Mark Dionne, Kevin K. He, Craig Hilton-Taylor, Charles Latrémouille, John Morrison, Calla V. Raymond, Mary Seddon, Varsha SureshA Standard Lexicon for Biodiversity Conservation: Unified Classifications of Threats and Actions, Volume22, Issue4, August 2008, Pages 897-911.
19. Enquist, BJ *et al.,* 2019: The commonness of rarity: Global and future distribution of rarity across land plants. Science Advances, 5(11), eaaz0414, doi: 10.1126/sciadv.aaz0414.
20. Brown, S. C. *et al.,* 2020: Persistent Quaternary climate refugia are hospices for biodiversity in the Anthropocene. Nature Climate Change, 10(3), 244–248, doi: 10.1038/s41558-019-0682-7
21. Trew, B. T. and I. M. D. Maclean, 2021: Vulnerability of global biodiversity hotspots to climate change. Global Ecology and Biogeography, 30(4), 768– 783, doi: 10.1111/geb.13272.
22. https: //epgp.inflibnet.ac.in/epgpdata/uploads/epgp_content/S000014ER/P000273/M027570/ET/1519296916Paper03_Module27_etext.pdf
23. Agrawal,K.C.(2009). Biodiversity: Concept, Conservation and Management. ISBN: 81-89153-05-6.
24. Kotwal, P.C. and Banerjee Sujoy (1988). Biodiversity conser-vation in forest and protected areas: Practical problem and prospects. In: Biodiversity of conservation, Agro Botanica. Bikaner
25. Shiva V (1994) Agriculture and food production. UNESCO/Environmental Education Dossiers No. 9 (May): 2–3
26. Alexandre Antonelli, Alexander Zizka, Fernanda Antunes Carvalho, Ruud Scharna, Christine D. Bacona, Daniele Silvestro and Fabien L. Condamine, Amazonia is the primary source of Neotropical biodiversity. vol. 115, no. 23, 6034–6039, PNAS, June 5, 2018

Chapter 4

Challenges to Biodiversity Conservation in the Twenty First Century

G. Sahaya Anthony Xavier

Associate Professor, Department of Botany, St. Xavier's College (Autonomous), Palayamkottai–627002
e-mail: saxsxc@gmail.com

ABSTRACT

The earth is the only planet on which life is known to exist so far. There are various forms of life on the planet. Life on earth is characterised by its diversity of form, structure and function. This variety of life called biodiversity is under threat due to various anthropogenic factors. Man is a product of life on earth. He cannot live in isolation. The dangers to biodiversity are ultimately a danger to the survival of the human species itself. Biodiversity conservation is essential to protect life as we know on the planet. There are many challenges to biodiversity conservation but there are also means of mitigating the challenges. A concerted effort needs to be mad to conserve biodiversity and save the planet for the danger of species extinction.

Keywords: *Biodiversity, Conservation, Evolution, Extinction.*

1. Introduction

The earth is home to millions of microorganisms, plants and animals. These living forms have evolved over a period of millions of years. Although all living organisms share the same type of genetic material, they have different forms, shapes and sizes. This variety of living forms is known as biodiversity.

The different forms of biodiversity are species diversity, which is the diversity of the different types of species living on the planet, ecosystem diversity which is the diversity of the different types of ecosystems present on earth and genetic diversity which is the diversity due to the differences in the genetic material *i.e.* DNA in the living organisms (Ashok, 2020).

The diversity on earth has not risen in a single day. It is the result of the eons of evolution that has happened till this day. In fact, even now, the processes of evolution are slowly at play although the processes are very slow to be visible during the life time of one human individual. As organisms adapt to the changing conditions or earth, as they compete for resources and mates, they slowly undergo changes due to random changes in their DNA. These changes in DNA slowly build up and over a period of tens of thousands of years, the organism undergoes a change, leading to new forms and species. Of course, there are organisms which seem to have not undergone any change in millions of years. These are called "living fossils". The examples of such organisms are *Ginkgo biloba*, commonly known as ginkgo, also known as the maiden hair tree which is a gymnosperm tree species found in South East Asia.

Every species that has arisen on the planet has a certain period of time during which lives on the planet. Just as an individual organism, whether microbe, plant or animal is born, grows, lives and finally dies, so to the species also has a timeline. New species are formed, they may grow and flourish. They may be present in small niche environments or they may be very well adapted to grow in a wide area on the surface of the earth. Over the period of millions of years, so long as the species is well adapted to its environment, the species will survive. If the global climatic conditions change, then the species who are unable to adapt to the changing environment, find that they are out of sync with the conditions on the planet. Such species may become restricted to very small areas on the globe where the climate is still favourable to their growth. Plants that are restricted to very small areas on the planet are called as the paleoendemics. They are relict plants which are past their heyday. Unless, the climate on the planet changes to their advantage, they will be present in their restricted habitats. If the climate further deteriorates, these organisms will become extinct.

Or, a species may have adapted well to the changing conditions on the planet and in due course it may have mutated or evolved into a new species. The old species is lost and becomes extinct. This process of formation of new species and extinction of species past their prime is a feature of the geological history of the earth. The fossil record of the planet shows us that there have been countless plants and animals in the past who are the ancestors of today's plants and animals but which are not living on the planet. This is a continuous and inevitable feature of life on the planet. Species are not infallible. New species are formed and old species go extinct. This phenomenon is as old as life itself and the process can be expected to continue into the future till the day when our sun itself will burn out all its fuel. At that point of time, which scientists estimate will happen some five billion years into the future, the sun will transform itself into a hot red giant. It will expand in size and engulf the planets Mercury, Venus and possibly earth itself. The temperatures on earth will increase so much that all the water will evaporate and the planet will heat up so that there would be no chance for any living organism to survive on the planet. Ultimately, our planet is doomed to end its life as a hot furnace due to the intense radiation from the sun and will end up as a life less rock, some five billion years in the future.

Life, however, has never had a smooth trajectory on this planet. While we may worry about some doomsday scenario in the future, the reality is that for most species on the planet, everyday life is a struggle. Life is threatened in various ways and living organisms have to invent strategies to survive and make this planet their own. There have been five major extinction events in the past, in the geological history of the earth. During these mega extinction events, the majority of life forms on the planet have been killed due to various reasons (Cherly and Junofy, 2012).

The mega extinction events on the planet have been as follows;

1. The Ordovician-Silurian extinction – this happened some 443 million years ago. It was most likely a fallout of global cooling, glaciation and falling sea levels on the planet. It is estimated that around 85 per cent of all life forms on the planet went extinct due to this event.
2. The Late Devonian extinction - around 375 – 360 million years ago, multiple events due to possibly climate change, oceanic oxygen depletion or asteroid impacts wiped out around 70 – 80 per cent of species on the planet.
3. The Permian – Triassic extinction – some 252 million years ago, massive volcanic activity led to global warming and ocean acidification. This led to the extinction of 90 per cent of aquatic species and 70 per cent of all terrestrial species. This event is known as the "Great Dying".
4. The Triassic – Jurassic extinction – volcanic activity some 201 million years ago led to global warming and climate change. This time too around 70 per cent of all living species went extinct.
5. The Cretaceous - Palaeogene extinction – it happened around 66 million years ago. Massive volcanic activity leading to the formation of the Deccan Traps in Central India along with the impact of the asteroid, which formed the Chicxulub crater are believed to be responsible for this extinction event. This event led to the extinction of 75 per cent of the species on the planet (Gabriel *et al.,* 2010).

One common thread to note among all the above mega extinction events is that all the extinction events have led to massive death of the species. The biodiversity on the planet plummeted. Life has however, recovered all the five times. The species that survived the extinction event were able to diversify and fill all the ecological niches available on the planet.

We are now believed to be in the midst of the sixth mega extinction event. Unlike the previous five extinction events, which are believed to be due to natural causes, the responsibility for the current extinction event lies solely on our shoulders. The speed at which the current extinction event is taking place is also astonishing and incomparable for the effect one species can have on the various life forms of the planet. If we consider the start of the Industrial revolution circa 1750 as the starting point at which our activities start to hamper and damage the planet, then it is remarkable that in a short span of 275 years, we have been able to disturb the balance of nature that has been forged over millions of years of evolution and co-adaptation (Pievani, 2014).

To compound the problems of biodiversity loss, it is not as if all the members of the human race are working together as one community. The human species is distributed around the planet and our numbers exceed 8.2 billion today as per estimates provided by the United Nations. With a global growth rate of 0.85 per cent annually, we are adding 70 million people every year. These many people are living in different countries of the world, numbering around 200 at present. They speak multiple languages, follow different cultures and customs and their interests may not be aligned in sync with one another. Much of this population is present in two countries, India and China with India currently having the highest population on the planet. Close to 1.46 billion people live in India and 1.42 billion people live in China. The constraints of feeding such huge populations and providing them with a means of livelihood, clothing and shelter are huge to say the least. In such conditions, conserving nature or conserving the biodiversity in nature is the last thought on the minds of the leaders and the planners of the nation (United Nations, 2021).

The beginning of the industrial revolution heralded a dramatic shift in the way humans used the resources on the planet. There was a spike in the use of energy for various purposes. Machines were created that made more machines and more work was done in a day than had ever been done before. New methods of mining, agriculture and production came into vogue. There was a corresponding increase in the population. The needs of the population also increased in terms of food, clothing, shelter and transport.

The type of energy used before the Industrial Revolution was also different. It was primarily energy from wood or biomass and muscle for doing other work. Muscle power was provided by draught animals which again depended on green plants for their food. There was very limited use of wind power and water power.

The advent of the Industrial Revolution marked a transition from the use of energy from wood to coal, although this transition was very gradual at first. It was with the development of the steam engine by James Watt that made extensive use of coal necessary to fire the railway engines. The human society moved from a biomass based energy utilisation system to a coal based energy utilisation system. Therefore the Industrial Revolution is characterised by a rapid change in the use of coal as a source of energy (Stearns, 2013).

Combined with the increase in the human population, the search for new resources led to large scale extraction of resources from various parts of the planet. The advent of the railways led to the possibility of extracting resources and transporting it over large distances with speed and the costs came down when compared to the earlier methods of transport by animal wagons or ships.

All this together led to a concentration effort to extract resources from nature. The engines of industry needed more and more of different types of raw materials. New discoveries in science and technology facilitated the development of new materials and consequently new products. Science promised a dazzling future that was built on the extraction and exploitation of resources.

To be fair to those times, knowledge about the natural world and the inter-linkages between them were absent or still in its infancy. Although the ancients had used the plants and animals for various purposes since the dawn of humanity, the inter-linkages between plants and animals were unknown. The foundations of ecology were laid by Ernst Haeckel in 1866, which is a good hundred years and more after the start of the industrial revolution. The recognition of ecology as an independent subject worthy of study had to wait till the beginning of the twentieth century.

The two world wars fought in the beginning and middle of the twentieth century played a major role in destruction of forests and killing of wildlife. The establishment of plantations for various produce such as tea, coffee and spices played an important role in the conversion of native forests into plantations of various kinds. The destruction of nature for various other purposes found its echo among the learned men of various nations. The fast and non-forgiving nature of machines and the industrial production process, the settlement of workers in ghettos without any care for their comfort, health or wellbeing and the maximisation of profits as the sole motive guiding a business had its detractors.

The realisation that unfettered development and extraction of resources was harming the natural world and the human race was indulging in whole sale loot and plunder of the natural resources dawned on learned men and women in the 18th century. This came due to the awareness that species were becoming extinct due to human activity. Species extinction could not be denied not only because of the work of naturalists in documenting such episodes but also due to the realization that deforestation of large stretches of land and hunting were wiping out species from the face of the earth. This concern however did not reach the general public till the twentieth century.

Naturalists like Georges-Louis Leclerc and Comte de Buffon, raised the alarm bells as early as the 1750s that species were being wiped out in the North American continent and in the Caribbean due to human actions. Alexander von Humboldt documented as to how in South America, colonial agriculture disrupted ecosystems and he was able to link the destruction of ecosystems to human actions. The nineteenth century brought the issue of extinction of animals into greater focus.

The loss of the Dodo due to human actions, the loss of the great auks and the sharp decline of the passenger pigeon brought the question of loss of animals and plants, which are the products of millions of years of evolution closer to the public. There arose voices of conservation and the idea that the human race should act as a steward and protect the other organisms living on the face of the planet got hold.

As we have seen, one of the characteristic features of life on our planet is its variety and variability. This has arisen over the course of millions and millions of evolution. It has to be borne in mind that the five mega extinction events in the past have wiped out large swathes of the living world at different periods of time in the geological history of our planet. It should be borne in mind that it is estimated that only around 1 per cent of all the organisms that evolved on this planet are alive today. While organisms evolve and new species are borne all the time, although at a very slow pace for humans to see it directly during the lifetimes, the converse is also true. As new organisms evolve, older species become extinct. Although the calculation of the total number of organisms which have evolved on the planet and the number of species that have gone extinct includes doing a lot of guess work, it is safe to assume that due to the natural rate of extinction of species and the five major extinction events on earth, the majority of species that lived on the planet are now extinct.

The term biodiversity was coined by Walter G. Rosen in 1985 but the idea it represented had gained a foothold in the minds of the people involved various aspects of wildlife protection. Rachel Carlson in her book, "The Silent Spring" published in 1962, had exposed the dangers of DDT pollution in the food chain. The book had made it amply clear the effect DDT had on non-target organisms. Moreover, the book clearly demonstrated that all the organisms and ecosystems on the planet are intertwined with one another and any disruption or harm in one part of the planet would definitely lead to harm elsewhere.

Increased research in the life sciences and ecological sciences showed us that the planet is not a series of dissimilar and disconnected ecosystems but all the ecosystems are connected as a gigantic whole. The Gaia hypothesis was put forward which considered the planet to be a giant organism which could regulate itself. Studies on the various forces operating on the planet like ocean currents, tidal waves, the biogeochemical cycles all showed that all parts of the planet were connected (Lovelock and Margulis, 1974). We who live in India depend on the monsoon for our water needs. Indeed much of India gets its rain during the four monsoon months from June to September and almost little or no rain during the rest of the year. In Tamil Nadu, we get rain from the North-East monsoon or the retreating monsoon. This occurs in a very narrow window between October and mid-January. Any disruption or delay in the onset of the monsoon affects the lives of all the people and the living organisms that live on the Indian subcontinent. It is important to realise that the monsoon originates near Madagascar in the Southern Indian Ocean. Therefore, any activity that affects the movement of air in the Southern Hemisphere such as the El Nino or the La Nina, affects rainfall in the Indian subcontinent (Vikas and Dwarakish, 2015).

The human species on the planet cannot survive in isolation but only together with all the other species living on the planet. All the other species on the planet from the plants to the animals form a complex web or net of which we are only a miniscule part. The planet can survive without the human species but the human species cannot survive with the myriad numbers of plants, insects, animals and microorganisms which together harbour and support the web of life.

The realisation that biodiversity is a key feature of life on this planet and that we have to preserve and protect this biodiversity not for some altruistic motive but for our survival has led to many efforts at biodiversity conservation. Biodiversity conservation is the practice of protecting and ensuring the survival of different forms of life on the planet. This encompasses species, ecosystems and the genetic diversity of life on the planet. The biodiversity is ensured not only for us but also for the future generations. This is partly also due to the realisation that the earth does not belong to us humans alone, but all the organisms living on this planet call it their home and they have a right to exist and coexist with us and other organisms.

The question may arise as to what would happen to a planet without biodiversity? All living organisms on the planet are able to survive due to the evolution of life over the past 3.6 billion years has given rise to various ecosystems which support the organisms. No living organism is greater that the ecosystem that it finds itself in. A planet without biodiversity would lead to ecological collapse. The collapse of the ecosystem would render the planet uninhabitable for most life forms on earth including humans. For example, we rely on a number of plants for our food, although the primary forms of food we consume are the cereals. However, the plans themselves rely on a number of insects to act as pollinators and ensure the dispersal of the pollen grains from the anther to the stigma. Without the insects, many plant species will not be able to transmit their genes to the next generation. Adding to the complexity is the fact that not all insects can pollinate all species of plants. There exist many specific relationships between the flowering plants and the pollinator species. So a particular species can pollinate only a specific flowering plant. The death of the pollinator will mean that the species which it has specialised to pollinate will also suffer and unless it has other insect species pollinating it, the plant species will also go extinct.

Some insect species are generalised pollinators and they have a very wide area of influence. *E.g.* the honey bees pollinate many species. It is estimated that if there would be no honey bees to pollinate the flowers, the food security of humans would be in jeopardy. So for humans to get a continued supply of food, not only do we need the plants to prepare our food, but we also need the insect populations that pollinate the plants to be healthy too. In other words, man depends on the insects which he many consider to be a low form of life for his survival. That is a humbling thought.

Ecosystems provide us many ecosystem services about which we may not give a thought and take the availability of the resource for granted. The availability of oxygen in the atmosphere, pollination, water purification, decomposition of dead organisms, the cycling of nutrients in the atmosphere are all ecosystem functions for which nature does not send us a monthly bill. Were we to attempt the same using human systems, the cost in terms of material, energy and manpower would be too great for us to pay it. All the above critical processes and supported by biodiversity and hence for a healthy planet, the biodiversity of the planet needs to be preserved. Indeed, a healthy planet is a biodiversity rich planet. Biodiversity conservation aims to preserve the intricate web of life itself, from the smallest microbes to the largest mammals and the ecosystems they depend on for their survival and sustenance.

Biodiversity conservation is primarily concerned with two goals,

a. to preserve the richness of species on the planet
b. to preserve the ecosystems that support the various species.

By fulfilling the above two goals, the third goal of biodiversity namely to protect and preserve the genetic diversity of species on the planet will be fulfilled. Preserving the genetic diversity of organisms is important as species having a diverse gene pool will be able to face the changes happening on the planet. Genes are the raw materials which provide organisms with the resilience to adapt to climate change, fight disease and survive in the face of an ever changing environment. The aim of biodiversity conservation therefore is not only the conservation of a species but to also ensure the genetic fitness of a species so that the species can propagate itself into the future that is uncertain.

It therefore becomes very clear that for the survival of the human species, the biodiversity needs to be conserved. Hence biodiversity conservation is of paramount importance. However, conserving the biodiversity of the planet has always been a challenge. It is more so today due to the following factors,

2. Population Explosion

Malthus postulated a theory about population growth and resource limits. He envisaged a scenario wherein the population of humans on the planet would rapidly multiply and overtake the ability to produce food for all human beings. That would lead to checks on the population numbers and bring the population down. This was almost a doomsday like scenario. However, although our numbers have crossed eight billion and are nearing nine billion, the doomsday scenario foreseen by Malthus has not yet fully come to pass. Using technological innovations, especially in the field of agriculture, such as the development of hybrid crops, chemical fertilizers and pesticides, and other types of innovation, mankind has been able to produce food to sustain its population. It is also pertinent to note that this increase in agricultural production has also been possible due to the conversion of large tracts for forest land to agriculture. A very good example of the same has been the conversion of large tracts of old world rainforests in parts of South East Asia for the cultivation of palm trees (*Elaeis guinanesis*) for the production of palm oil. Palm oil is cheap and it is used widely for cooking. The real costs involved in the use of palm oil include the destruction of the rainforests and the loss of the wildlife and the valuable gene pool due to the loss of biodiversity. Incidentally it makes no difference to the people in government and in the private companies cultivating the palm oil trees on forest land that these forests are home to a number of rare and endangered species including the iconic Orang Utang – which are among the great apes and very close to us in the evolutionary family tree.

Thus, so far, the doomsday predictions of Malthus have not come true. However some experts point out, that even today there are huge pockets of malnutrition and silent hunger in many parts of the world. In India, it is estimated by FAO in 2023 that 74.1 per cent of Indians are malnourished and 16.6 per cent Indians are under nourished. 31.75 per cent if children under five in India are stunted and 53 per cent of women between the ages of 15 – 49 are anaemic. Similarly there are many countries in South Asia and Sub-Saharan Africa where hunger in endemic. The FAO report on "State of Food Security and Nutrition in the World" in its 2024 edition states that 733 million people were undernourished globally which is 9.1 per cent of the world's population. With the human population projected by the United Nations to hit 9.6 billion by 2050 and reach 10.3 billion in the 2080s it is a challenge that is waiting to be met. Let us hope that Malthus may not be proved right.

3. Habitat Loss

The conversion of natural environments, not only forests but also marsh lands, wet lands, salt pans, water bodies, dam building, industrial mining and a host of other factors including clearing land for agriculture lead to habitat loss. Habitat loss is detrimental not only because the plants and animals living there are affected. When entire habitats are destroyed, the ecosystem processes functioning there are also destroyed. Unfortunately, the natural environments such as forests have vast reserves beneath them such as coal, tar sands, precious minerals, *etc.*

The demands of industry are huge. It is often not possible for the small natural tribal communities which live in these natural habitats to stand up to the might of the corporations. All around the globe from the Amazon rainforests, to the Congo rainforests, to the forests of South Asia and in India, indigenous communities are on the war path to protect their forests and with it their way of life. Often times they are not successful.

Habitat loss destroys entire ecosystems. The intricate and delicate balance between the organisms living in the particular habitat is destroyed and in most cases, it may never be possible to repair the damage. Even if damage is attempted, it will take a long time before things may reach a state of balance again.

4. Fragmentation of Forests

The building of road and dams within forests leads to fragmentation of forests. Fragmentation of forests affects the animals living within them because the animals need to move from feeding grounds to nesting grounds *etc.* The bigger the animal, the bigger the problem. Animals do not respect national boundaries. Elephants are known to walk around the African continent round the year to find sources of food and water. They are constantly on the move as they require huge volumes of food and water to sustain them. They have been doing so for thousands of years but now humans have divided the land between them into nation states and sometimes put barriers to their movement. Thus fragmentation of forests leads to escalation of human – animal conflict primarily because humans are going into environments where they have not gone before.

5. Global Warming and Climate Change

The current event of global warming is a manmade disaster. The burning of fossil fuels in huge quantities to satiate our needs has resulted in the increase in the atmospheric carbon dioxide concentration from around 0.25 per cent (250 ppm) in 1750s at the start of the industrial revolution to well past 0.04 per cent (400 ppm) today. It is pertinent to note that the concentration of oxygen in the atmosphere is 20.6 per cent and it does not do much harm to the living organisms. Carbon dioxide is a greenhouse gas that causes warming of the atmosphere. Consequently a small increase of around 150 ppm of carbon dioxide over a period of 275 years has put the future of humanity at jeopardy. There is no doubt that life will adapt to the increase in the carbon dioxide concentration over the course of thousands or millions of years. The question is whether the human species will be able to survive the effects of global warming and climate change.

6. Pollution

Industrial development has led to different types of pollution such as water pollution, air pollution, soil pollution, radiation pollution *etc.* The emission of gases like sulphur dioxide, nitrogen oxides and particulate matter mix with water vapour in the air forming acid rain or smog. Acid rain on falling to the ground increases the acidity of the soil and the water bodies. It kills the plants, fish and amphibians which are sensitive changes of pH.

The pollution of water bodies due to uncontrolled use of chemical fertilizers, pesticides, heavy metals *etc.*, leads to eutrophication of water bodies. This leads to decrease in biodiversity of natural aquatic systems.

Similarly soil pollution, radiation pollution and thermal pollution also lead to change in the ecosystems and lead to decrease in biodiversity.

7. Overexploitation

We use many species for different purposes not only as sources of food. However every species has its own birth rate and growth rate. Overexploitation refers to the situation where species are harvested much faster than they can replenish themselves. This drives down biodiversity by reducing populations much faster than they can reproduce. Some species are driven to extinction, others may face a population bottleneck and hence may not be able to bounce back to healthy numbers. In the case of keystone species, the attack on a single species has ripple effects that are felt through the ecosystem. For example, elephants are a keystone species on which the health of the ecosystem depends. If elephants are hunted to a limit where their population numbers are down and they are not in good numbers to play their role in the ecosystem, then the ecosystem will suffer.

Overexploitation leading to the collapse of wild populations has been seen in the case of Atlantic cod (*Gadus morhua).* Over 99 per cent of the species was harvested leading to a population crash. There were too little fish to catch. Consequently a moratorium on fishing was enforced. It led to many fishermen losing their livelihoods and facing hardships. When the population of the cod fell down, the species which it held in check like the shrimps and crabs proliferated, which then overgrazed. The resulting cascade shifted the entire marine ecosystem, diversity plummeted. Overfishing of the Atlantic cod affected not only the cod but the entire marine ecosystem.

Such a situation has been seen in the case of many medicinal plants. Overharvesting the plants such as ginseng (*Panax ginseng*) and Himalayan yew (*Taxus baccata*) for medicinal purposes has mean that the population has crashed in many places. As seen earlier, plants share a relationship with other insect pollinators. Therefore the crash of plant populations, leads to loss in diversity of other species also.

The Asiatic lion was once prevalent all over Asia and the Middle East. Bas reliefs from ancient Assyria show their king Ashurnipal fighting a lion and subduing it. Today the entire population of the Asiatic lion is reduced to the Gir forests of Gujarat. Around 300 lions remain and if something catastrophic were to befall the lion population of Gir, such as an epidemic, we stand to lose an iconic species. This is the result of centuries of hunting the lion.

8. Problem of Invasive Species

Global travel is something that is taken for granted. From the massive passenger ships to aeroplanes criss crossing the planet, today travel is one of the great accomplishments of the human race. Our cargo ships keep moving round the planet, transporting men and material and making modern life easier. Along with the cargo ships and aeroplanes, plants and animals also move. Consequently we have the problem of invasive species, species which are out of their native range but which have now expanded to a completely new ecosystem. The invasive species very often do not have any predators to keep them in check in the new environment in which they find themselves. As a result they multiply quickly and overrun the ecosystem. There are many plants and animals which are invasive species and which have become problematic around the world. Plants like *Eicchornia crassipes, Prosopis juliflora*, kudzu (*Pueraria Montana*), Japanese knotweed (*Fallopia* japonica), animals such as the Indian crow (*Corvus splendens*), the European starling (*Sturnus vulgaris*) and the Burmese python (*Python bivittatus*) are all invasive species that are wreaking havoc in the new environment where they find themselves.

The invasive species have severe ecological effects. They threaten local species which in many cases may have no defence mechanisms. Feral pigs are known to threaten the native forest birds in Hawaii and are known to cause species extinctions especially of endemic species. In islands like New Zealand or the Galapagos islands, the invasive species are the leading cause of species extinctions. The Convention on Biological Diversity has identified the invasive species as one of the leading causes of species extinctions and biodiversity loss worldwide alongside habitat destruction and climate change.

Fighting invasive species is a long term challenge that will need the governments and scientists to work together. Although it is a difficult task, success stories like the eradication of rats from certain Pacific islands to save sea birds show that it is possible to eradicate invasive organisms. The invasive organisms need to be tackled for if left uncontrolled they will threaten ecosystems, economies and human health on a large scale.

9. Economic Activities

Economic activities affect biodiversity often depending on the way it is practised. The clearing of natural areas for various economic activities such as farming and mining leads to loss of pristine lands which leads to loss of biodiversity. Urbanisation, industrialisation, mining for minerals, logging, manufacturing, drilling for oil or mining for coal to drive energy production, conversion of natural ecosystems for ranching are some of the economic activities that drive habitat loss and loss of biodiversity. Industrial fishing has depleted fish stocks around the world.

It is however, not necessary that economic activities should lead to biodiversity loss. Carefully planned development keeping the needs of the environment can reduce if not complete reverse the losses due to economic activities. Farming practices such as organic farming or agroforestry can ensure better biodiversity across farmlands. In Costa Rica, coffee plantations incorporate shade trees to protect bird biodiversity. Similarly, instead of hunting animals for pleasure, ecotourism activities generate revenue by allowing visitors to visit the natural parks and see the wild animals in their natural environs. Timber companies are mandated by law to replant the areas they have cut down for timber. The adoption of green technologies can mitigate the losses due to economic activities. China has implemented the Grain for Green program where farmers are paid to convert marginal or sloping farmland into forests and grasslands to combat deforestation, soil erosion and to improve the general ecological conditions of the area. These and other steps show that when there is a will on the part of the governments and the people, they can work together to protect biodiversity.

10. Technological Advances

The advances in technology today are mind boggling. There is scarcely any person on the planet who today would not be affected by the march of technology. Technology has made people's lives easier. The availability of smart phones and internet data packs at an affordable rates, has transformed society. The world is present at the touch of the phone button and most of our purchases, bank transactions, financial transactions, *etc.* can all be done in the comfort of our homes.

However, when technology is not used properly, it has its pitfalls. The largest organism to have ever lived on the planet lives in our seas. It is the blue whale which can grow up to 100 feet and weigh 200 tons. Technological advances mean that we can now build big ships which can use modern technological tools like sonar to hunt the whales. The entire whale can be captured, hauled onto the big ships, cut into pieces and the meat stored in the gigantic refrigerators aboard the whaling ships. Our technological advances have meant that whale populations have plummet. It is only now that after the International Whaling Commission has placed restrictions on the hunting of whales that some populations are showing signs of recovery. It should be noted that whales are giant animals and they reproduce very slowly when compared to some of the smaller animals.

Similarly, our quest for having better and better smart phones and computers and electric cars have led to a global search for rare earth minerals and elements. No part of the planet is free from the search for rare earth minerals. There are reports of such minerals being mined in the rainforests of the Congo much of which is protected. Companies now want access to mine the deep sea floor which is believed to have many rare earth minerals. We now have the technology to mine the ocean floor. We do not have a proper knowledge of the various organisms and ecosystems inhabiting the ocean floor. In such a scenario, does one species have the right to mine for minerals and destroy the countless communities living on the ocean floor?

11. Cultural Practices

Human societies have culture and the culture among the various communities varies. In many forest dwelling communities, there is the practice of slash and burn culture, wherein a piece of the forest is cut down and burnt to practise agriculture. Generally such plots of land do not remain fertile for a long period of time and hence the plots of land are abandoned after a few years and the forest community would move to another area and repeat the process. When the population of forest communities was less, the plot would not be reused by the forest community for up to twenty or thirty years. This would give ample time for the forest to reclaim the area and the plot would return to its natural state. Today the population of the communities living in the forest too has increased. This means that an abandoned plot is reclaimed for agriculture within a few years of it being abandoned. This means that there would be no nutrition in the soils to support the crops and the agriculture becomes a failure. The people depending on such a form of agriculture faced great hardship.

Hunting is considered as an ancient way of life and a source of pride by such communities. They may resist attempts by the governments to regulate or to ban hunting. The hunting of wild animals for their flesh, called bush meat, is prevalent across many regions of the world. Almost all the animals in the forest are fair game. The bush meat trade is driven by poverty of the people who see in it a means to earn some money. Here poverty leads to loss of biodiversity.

12. Global Trade in Exotic Plants and Animals

All around the world there are enthusiasts who want to grow exotic plants and animals in their homes. The issue is well documented in many countries of the world. In the United States of America, the Burmese pythons that are now wrecking the Everglades ecosystem in the State of Florida were first imported into the country to be grown as pets. When the pets became too big to be taken care of in the houses of people, they were let loose in the wild. Having no predator to control it, the Burmese python took over the ecosystem and is now busy reengineering the ecosystem to its advantage.

Almost all sorts of animals and birds that have some aesthetic value are sought to be traded in. There are international trade agreements like CITES (Convention on International Trade in Endangered Species) but there exist networks to bypass the law enforcing agencies in the different countries of the world. The dark net is said to be populated by such elements who are dealing with the murkier aspects of trade.

These are some aspects of the challenges in biodiversity conservation. There are many more. Enlisting them all would require a big book to be written. The question is, "What can be done to conserve the biodiversity"?. The first of the steps that need to be taken is,

a. Population control – the earth can only feed and support a given number of people. There is no consensus on what is the number of people that can be supported by the planet. There is no doubt however that the numbers we now have in excess of eight billion and more cannot be sustained for a long period of time.

b. Poverty eradication – it is the poor who depend on biodiversity for their survival. In the fight against loss of biodiversity, the poor must be enlisted as foot soldiers. The people living in the forest could be employed as forest guards, as watchers and forest guides. By ensuring that the people get livelihood by protecting the forests or natural preserves, the message that forests should be protected would get across to the people. The people would not feel alienated from the forests.

c. Combat climate change - climate change remains one of the greatest dangers to biodiversity conservation. Climate change threatens to overthrow the very fabric of life on our planet. Combating climate change requires decision making at the highest level and international cooperation between all nations. Just as the world came together to fight depletion of the ozone layer by forming the Montreal Protocol, we need another round of statesmanship to combat climate change.

d. Control Invasive Species – international trade in endangered animals and plants should be completely eradicated. The organisations involved in fighting invasive species should be given more powers so that this threat can be eliminated.

c. Education – education is the key to fight for biodiversity conservation. Every child and every adult must be reached. All citizens of the planet should know about the importance of biodiversity and the measures that need to be taken to protect it.

d. Better use of resources – all the nations of the world spend very heavily on their defence budget. We are perhaps a very unique species who spend a lot of time, energy, money and resources to devise more and more effective ways to kill one another. If the world could achieve world peace, if all the nations of the world could learn to live in peace and harmony, then much of the defence budget could be used to solve the problems caused by loss of biodiversity. There would be ample funds for afforestation, for green chemistry, for poverty alleviation and search for alternate materials and resources which could propel development without compromising biodiversity.

13. Conclusion

Our world is at a cross roads. As stewards of the planet and as the one species chiefly responsible for the problems of biodiversity, it is primarily our responsibility to ensure that we solve these problems. We should learn to share the earth with other species. We have to realise that the steps we take to conserve biodiversity are in reality steps to save our species. A biodiversity rich planet could guarantee a secure future for our future generations.

14. References

1. Ashok Kumar Verma, 2020. Biodiversity: its different levels and values. *International Journal on Environmental Studies*, 7(2): 143 -145.
2. Carson, Rachel. 1972. The Silent Spring. Boston: Houghton Mifflin.
3. Cherly Anto Frezina, N and Junofy Anto Rozarina, N. 2012. Extinction of species and its management for ecological balance. *Journal of Educational and Social Research*, 2(9): 159-163.
4. Gabriel Dorado, Isabel Rey, Francisco Javier S, Sánchez-Cañete, Fernando Luque, Inmaculada Jiménez, Arturo Morales, Manuel Gálvez, Jesús Sáiz, Adela Sánchez, Pilar Hernández, Víctor F. Vásquez. 2010. Biological Mass extinctions on planet earth. *Revista Archaeobios*, 4(1), 53-65.
5. Lovelock, E. James and Margulis, Lynn. 1974. Atmospheric homeostasis for and by the biosphere: The Gaia Hypothesis. *Tellus*, XXVI, 1-2.
6. Pievani, Telmo. 2014. The sixth mass extinction: Anthropocene and the human impact on biodiversity. RENDICONTI LINCEI. 25. 10.1007/s12210-013-0258-9.
7. United Nations Department of Economic and Social Affairs, Population Division, 2021. *Global Population Growth and Sustainable Development.* UN DESA/POP/2021/TR/NO. 2.

8. Stearns, N. Peter. 2013. The Industrial Revolution in World History, 4th Edition, Westview Press, Colorado.
9. Vikas, M and Dwarakish, G. S. 2015. El Nino: A Review. International Journal of Earth Sciences Engineering, 8(2): 132-137.

Chapter 5

Ethnobotany

J.L. Lekshmi

Assistant Professor, Department of Botany, S.T. Hindu College Nagercoil, Affiliated to MS University, Tirunelveli– 629002, Tamil Nadu
e-mail: mathulekshmi@gmail.com

ABSTRACT

In broad terms, ethnobotany is the study of the interrelationship between plants and people. The two major parts of ethnobotany are encapsulated in the word itself; ethno, "the study of people", and botany, "the study of plants". However, the field is limited on both sides. It may be defined as the study of a region's plants and their practical uses through the traditional knowledge of a local culture and people. The field combines botany, anthropology, economics, and sometimes archaeology (in paleoethnobotany, the study of vanished agricultural systems). The nature of relationships between man and plants however, may be either Abstract (worships/rituals, faiths and beliefs, natural powers, taboos, sacred plants, folklore, legends, sylvan environments, myths, etc.) or Concrete (material use, cultural use, conservation, disuse).

Keywords: Ethnobotany, Anthropology, Agricultural systems, Sacred plants, Conservation

1 Introduction

The prehistory man must have used plants as food and for curing ailments. A variety of uses of plants are mentioned in the ancient Indian Sanskrit literature like Rigveda (ca 4000-1500 BC), Atharvaveda (1500 BC), Upanishads (1000-600 BC), Mahabharata and Puranas (700-400 BC), *etc.* These include plants used in worships, as medicines, food, fuel and for tools of agriculture. A specific mention may be made of the uses of different varieties of Rudraksha (Elaeocarpus ganitrus) in Lakshmipuran and Shivapuran. Uses of about 1200 plant drugs along with their action and specific therapeutic applications are mentioned in Sushruta Samhita (ca 500 BC), Charaka Samhita (ca 100 AD) and Ashtanga Hridaya Samhita. In fact, many more references of plants are reported but their identity has not yet been established. Richard Evans Schultes, often referred to as the "Father of Ethnobotany". Dr. Sudhanshu Kumar Jain [Dr. S. K. Jain] (NBRI, Lucknow) affectionately known as "Father of Indian Ethnobotany"

2. Ethnobotany as an Interdisciplinary Science

2.1. Interdisciplines of Ethnobotany

Ethnopharmacology deals with the identification, description, observations and experimental investigations of the ingredients used in various recipes prepared by aborigines, and the effects ofindigenous drugs on laboratory animals and man.

Ethnomedicine includes research that deals with medicines derived from plants, animals, minerals, *etc.*, and used in the treatment of various diseases and ailments, based on indigenous pharmacopoeia, folklore and herbal charms.

Ethnotoxicology deals with the use of various toxic plants as fish poison (Ichthyotoxic), arrow poisons, *etc.*, in human societies.

Ethnomusicology is the study of music of tribals and aborigines, its documentation, forms and content.

Archaeoethnobotany involves three subjects namely, ethnology, archaeology and botany. This interdiscipline of ethnobotany studies the identification of plant materials from archaeological sites for studies on migration of human cultures, and origin, dispersal and domestication of crops, *etc.*

Ethnogynaecology is an emerging discipline that deals with various diseases among women in tribal societies, related to sterility, conception, abortion, *etc.*, and use of abortifacients.

Ethnonarcotics is the study of the use of narcotics, snuffs, hallucinogens, *etc.*, in primitive societies.

Ethnogastrology is the study of ethnic groups in relation with their food and food habits. It also includes traditional recipes of different ethnic groups.

Ethnopediatrics involves study of the science of health care in children practiced among different ethnic groups.

Ethnoorthopaedics is the study of knowledge of different ethnic tribes/ communities about setting and healing of bones.

Ethnohorticulture involves study of management of useful plants (*e.g.*, fruits, vegetables, ornamentals) in home gardens or orchards.

Ethnoforestry deals with the study of human management of forests and forest trees.

Ethnoagroforestry is the study of land management for the simultaneous production of food, crops and trees.

Ethnolinguistics is the study of linguistic terminology for plants and plant parts by people of different language groups.

Ethnosilviculture deals with practice of raring plants of economic and cultural values. Ethnozoology is the study of complex relationships between people and animals, which includes human relationships with and use of domesticated animals, and management of wild animals for hunting and other uses.

Ethnoentomology deals with the study of relationships between people and insects. This can include studies of useful insects and their products (honeybees, silkworms, *etc.*) and poisonous species as well, insects used in rituals, beliefs (taboos, avoidance), *etc.*, and how people interact with and control pests. Several such combinations of words could be possible where the knowledge of ethnology and various other science subjects is required. Today, ethnobotany is regarded as an independent subject and the prefix ethno is generally taken to indicate interdisciplinary nature of the subject. In view of the interdisciplinary nature of ethnobotany, an ethnobotanists should be familiar with at least six disciplines namely, botany, anthropology, chemistry, linguistics, economics and ecology, but should try to specialise in any one or two of them. The field of specialisation which an ethnobotanist may choose include traditional taxonomy, morphology, plant chemistry, drug plant cultivation, anthropology, traditional medicine, pharmacy, pharmacology and even therapeutics. A basic knowledge of floristics and anthropology is essential for an ethnobotanist.

Ethnobotany deals with not only the useful relationship between plants and man but also deals with the impact of man on his environment and vegetation. E. K. Janaki Ammal started Ethnobotanical study in India and often remembered as, "Mother Of Indian Ethnobotany". She studied the utility of food plants of certain tribals of South India. Dr Sudhanshu Kumar Jain is regarded as Father of Indian Ethnobotany. The Economic Botany deals with the uses of processed, improved or otherwise cultivated plant products and their Commerce; whereas Ethnobotany primarily deals with the uses of unknown or less known useful wild plants. According to RR Rao, Ethnobotany is a field oriented study involving direct relationship of aboriginals with surrounding plants. The man-plant relationship can be broadly classified into two categories, *viz.*

(a) Abstract Relationships: Includes faith in the good and bad powers of plants, taboos, avoidances, sacred plants, worship and folklore.

(b) Concrete Relationships: Includes mainly the material use such as in food, medicine, house building, agricultural operation, other domestic uses, plants in fine arts, and culture and acts of domestication, conservation, improvement, and destruction of plants.

Jain (1989) gave a brief description about all relationship between man and plants can be first divided into material and cultural relations, and placed in four categories:

1. Relationship useful to both man and plants [+/+].
2. Relationship useful to man but harmful to plants [+/-].
3. Relationships useful to plants but harmful to man [-/+].
4. Relationship harmful to both man and plants [-/-].

The other major aspects of ethnobotanical work are the findings of new uses or new economic plants which includes less known uses of plants like their use as fodder, fibre, fuel, oil, gum *etc.* Impact of ethnobotanical study in conservation of natural resources is direct. On one hand it relates to the beliefs, taboos, avoidance and other constructive approaches of the primitive people and on the other hand cultivation practices of some primitive people, adversely affecting the environment are highlighted. The aboriginals have preserved the forest through mythological associations as sacred groves.

Today ethnobotany has become an important and crucial area of research and developments in resource management, conservation of biodiversity at genetic, species and ecosystem levels, and socio-economic upliftment of the region. Many plants *viz.* Maize, Cocoa and Rubber used in industrialized countries today were originally identified and developed through Indigenous Knowledge [IK]. Traditional food plants such as roots and tubers (Cassava, Potatoes, Yams, Taro, Sweet potato and plantains nourish over 1000 million in the developing countries, providing a significant portion of essential calories to the rural people. Here mention may be made of a rare medicinal plant, *Trichopus zeylanicus* ssp. Travancoricus, used by the Kani tribe of Kerela, India as a health food to maintain vitality and increase resistance to disease. The tonic effect of the plant has been shown to be comparable to that of Korean ginseng (*Panax ginseng*).

The Winged Bean *(Psophocarpus tetragusonolob*), a crop native to South-East Asia, has great potential for feeding many in the Third World. One hectare of winged bean produces nourishment equal to five or six hectare of most other crops. Some of the so called 'New Crops' whose commercial production has been developed in the 20th century, such as African Oil Palm, Para Rubber, Robusta Coffee, Tepary Bean, Guayule Rubber and Jojoba were previously known and used locally by indigenous people. During the last few decades, successions of so called "Wonder Drugs (*e.g.* Reserpine, Quinine, Ephedrine, Cocaine, Emetin, Khellin, Colchicine, Digoxin, D-tubocurarine, Artimisinine and Gugullipid) have been discovered from plants with rich role in tribal societies. The Indian Tobacco (*Lobelia inflata*) used as a tobacco substitute by the Amerindians contains LOBELINE, now commercially used as a smoking deterrent. The May-apple (*Podophyllum peltatum*) has a toxic resin, PODOPHYLLOTOXIN formerly used by Amerindians to remove worts; its cytotoxin action has led to its modern use to treat uterine warts. Which formerly required surgery. The cardiotonic STROPHANTINE, comes from seeds of a species of *Strophanthus* that is used by African tribes as the source of its arrow poison. The tranquilizers, Rescinnamine and eserpine, have been obtained from the roots of *Rauwolfia serpentine*, used in India for more than a thousand years in folk medicine for snake bite, indsanity, epilepsy and high blood pressure. Similarly *R. vomitoria* of Africa and *R. tetraphylla* of America are found new source of Reserpine at commercial level.

The natives of Madagascar valued the rosy-periwinkle (*Catharanthus roseus*) as an oral hypoglycemic (Reduction of the sugar content of the blood) agent; it yielded two powerful drugs, VINBLASTIN and VINCRISTIN, effective against Hodgkin's Disease and Childhood Leukaemia. Hypertensive agents from Veratrum viride used by natives of America, Khellin from Ammi visnaga of the ancients, and many psychoactive agents developed from the Hallucinogenic plants of the New World Indians.

2.2. Four Major Interrelated Fields of Study in Ethnobotany

1. Basic Ethnobotany - Basic documentation of traditional Botanical knowledge.
2. Quantitative Ethnobotany - Quantitative evaluation of the use and management of Botanical Resources.
3. Experimental Ethnobotany - Experimental assessment of the benefits derived from plants, both for subsistence and for commercialization
4. Applied Ethnobotany - Applying traditional ecological knowledge to biodiversity

3. Conservation and Community Development

Numerous wild and cultivated plants play a very important and vital role among these cultures and this interrelationship has evolved over generations of experience and practice. These studies have therefore also been used in tracing human and plant migrations; origin, dispersal and domestication of cultivated plants like maize, tepary beans, squash, pumpkin, watermelon, bottle guard, jack bean, sword bean, common bean, foxtail millet, sunflower, amaranth, chenopod, sorghum *etc.*; in linguistic analysis; in archaeological identifications; agricultural techniques and agronomy, horticulture, pharmacopoeia *etc.* Practically all cultivated plants were originally domesticated by primitive people. The tribal belt is more often the centre of diversity and origin of our crop plants (Wheat, Rice, Maize). The ancestors of many of our crops and landraces are interwoven with tribal cultures.

The Ethnobotanical Literature has been growing rapidly over the last hundred years. Many important documents exist as Ethnographies, Archaeological accounts, Travelogues, Gazetteers, Herbals, Materia medica, Unpublished papers, Manuscripts and theses, and reports of the tribal development projects *etc.* intended for local distribution. There are number of medical manuscripts of traditional medicine lying in oriental libraries and private collections.

4. Indigenous Knowledge of Using Medicinal Plants in Treating Skin Diseases

Village communities from various ecosystems use largest proportion of biodiversity for human and veterinary healthcare.

Enumeration

The following plant species were identified for their use in traditional healthcare system against various skin diseases. The enumeration of plants contains botanical name, family name, local name and uses.

Acorus calamus Linn. (Araceae); Vasampu

Uses: Pounded rhizomes along with Curcuma aromatica rhizomes and Azadiracta indica leaves are applied twice a day, after bathing and before bedtime for one week against eczema.

Aegle marmelos (Linn.) Corr. (Rutaceae); Vilvam

Uses: Fruits crushed with seeds of Strychnos nuxvomica,Pongamia pinnata and boiled with coconut oil is applied on the affected parts to cure scabies and other similar skin diseases twice a day, till cured.

Anacardium occidentale Linn. (Anacardiaceae); Kollamaram

Uses: Powdered bark mixed with honey is taken orally once a day for six months against leprosy.

Andrographis paniculata Nees. (Acanthaceae); Nilavembu

Uses: Leaf juice mixed with cow milk is taken orally twice a day for six to eight days to cure tinea cruris.

Argemone mexicana Linn. (Papaveraceae); Premathandu

Uses: Pounded seeds along with rhizomes of Curcuma aromatica and Acorus calamus made into paste are applied on all kinds of skin diseases twice a day till recovery.

Asparagus racemosus Willd. (Liliaceae); Thannervittankizhangu

Uses: Tuber along with the leaves of Plumbago indica made into paste is applied on skin diseases once a day till cured.

Azadiracta indica A. Juss. (Meliaceae); Vembu

Uses: Flowers boiled in gingili oil (Sesamumindicum) is applied on head against dandruff once days in the morning after taking bath till recovery.

Cassia alata Linn. (Fabaceae); Seemai Agathi

Uses: Pounded leaves, coconut oil and bee waxmade into paste is applied on the affected parts to cure tinea versicularis once daily in night before bed time for four days.

Cassia auriculata Linn. (Caesalpiniaceae); Avarai

Uses: Dried leaf paste in vinegar is applied on skin diseases once a day till cured.

Clerodendron inerme Gaertn. (Verbenaceae); Changukuppy

Uses: Leaf juice mixed with bee wax, resins of Vateria indica and seeds of Nigella sativa, made into a paste is kept in a hot water bath and cooled before use. It is applied once a day before bed time till the recovery to cure various skin diseases.

Clitoria ternatea Linn. (Fabaceae); Shankupuspham

Uses: Leaf juice is given orally twice a day for six days to cure scabies.

Corallocarpus epigaeus Hk. f. (Cucurbitaceae); Kurudankizhangu

Uses: Tubers boiled in coconut oil is applied once a day for six months on the affected parts to cure leprosy.

Crinum defixum Ker. (Amaryllidaceae); Vishanarayani

Uses: Pounded bulbs mixed with hot water is given orally twice a day for three days for curing tinea cruris.

Curcuma aromatica Sal. (Zingiberaceae); Kasturimanjal

Uses: Rhizomes and Terminalia chebula seeds made into paste is applied on the affected parts twice a day till the recovery to cure impetigo.

Cynodon dactylon (Linn.) Pers. (Poaceae); Arukampullu

Uses: Pounded leaves boiled in coconut oil is applied once a day till the cure for various kinds of skin diseases.

Datura metel Linn. (Solanaceae); Ummattai

Uses: Leaf juice along with Curcuma aromatica rhizomes made into a paste is applied against swellings for fast relief till the swelling reduces.

Euphorbia hirta Linn. (Euphorbiaceae); Ammanpaccharisi

Uses: Latex is applied against skin parasites twice a day till cured.

Glycyrrhiza glabra Linn. (Fabaceae); *Atimaturam*

Uses: Paste prepared from pounded stem and Withania somnifera roots is applied on the affected parts once a day for one year to cure leukoderma and other skin diseases.

Hygrophila auriculata (Schum.) Heine. (Acanthaceae); Neermulli

Uses: Dried leaf powder mixed with castor oil is applied twice a day till the recovery on the affected parts to cure skin diseases.

Indigofera aspalathoides Vahl. (Fabaceae); Sivanarvembu

Uses: Powdered bark mixed with coconut oil is applied twice a day for six months on the affected parts to cure leprosy.

Lawsonia inermis Linn. (Lythraceae); Maruthani

Uses: Leaf paste is applied twice a day till it is cured on the affected parts to cure impetigo.

Madhuca longifolia (Koenig) Macbride. (Sapotaceae); Eluppai

Uses: Pounded seeds mixed with leaf extract of Ocimum tenuiflorum are applied on the affected parts twice a day to cure skin diseases.

Ocimum tenuiflorum Linn. (Lamiaceae); Tulasi

Uses: Leaves pounded with Curcuma aromatica rhizomes are applied on the affected parts once a day in the night before bed times to cure tinea versicularis.

Phyla nodiflora (Linn.) Greene. (Verbenaceae); Poduthalai

Uses: Paste prepared from leaf juice mixed with equal volume of gingili oil and boiled is applied twice a week on head to remove dandruff.

Piper betle Linn. (Piperaceae); Vettilai

Uses: Leaves pounded with Allium sativum bulbs is applied on the affected parts once a day in the morning after bath to cure tinea versicularis.

Pongamia pinnata (Linn.) Pierre. (Fabaceae); Punkumaram

Uses: Crushed bark paste boiled in gingili oil is applied on the affected parts twice a day for four days to cure rash.

Saraca asoca (Roxb.) de Wilde. (Caesalpiniaceae); *Asogam*

Uses: Paste prepared from dried flowers boiled with coconut oil and then cooled is applied thrice a day till cured with the help of cock feather against scabies.

Terminalia bellerica (Gaertn.) Roxb. (Combretaceae); Tanrikkai

Uses: Paste prepared from seeds pounded with the seeds of Terminalia chebula and Quercus infectoria and mixed with coconut oil is applied twice a day against rash.

Trichosanthes lobata Roxb. (Cucurbitaceae); Peppudal

Uses: Whole plant paste is applied once a day for one year on the affected parts to cure leprosy.

Wrightia tinctoria (Roxb.) R. Br. (Apocynaceae); Veppalai

Uses: Pounded leaves mixed with coconut oil are applied twice a day against psoriasis.

5. Medicinal Plants Used for the Treatment of Diabetes

The diabetes mellitus (DM) is one of the world's biggest healthcare challenges and it's increasing day by day, globally it affected about 171 million people in 2003, 173 million people in 2004, 284.6 million in 2009, 371 million 2012, while recently it is expected that up to 552 million people by 2030. Among all DM patients 90 per cent suffer from diabetes. About 80 per cent DM patients live in under developing countries, About 3.2 million deaths per year are caused due to complications of diabetes. The top most countries (in term of numbers) of DM patients are India, China, USA, Indonesia, Japan, Pakistan, Russia, Brazil, Italy and Bangladesh. The disease is caused by the inability of pancreas to produce insulin or inability of the body metabolic system to properly use the insulin produced. The cause of this disease are not known; however, it is suspected that occurrences of the disease may result from increased number of elderly people, change in food habits, obesity, and adoption of a sedentary lifestyle. Because the human body cannot properly metabolize sugar when during diabetes, the first symptom of the disease is an increase in blood glucose or hyperglycemia. The impaired metabolism of sugar and other dietary substances like lipids and proteins soon leads to disorders affecting almost all vital organs of the body. In fact, diabetic patients are extremely prone to develop neurological, cardiovascular, retinal and renal complications leading to premature death, which to a lot of extent may be due to diabetes-induced oxidative damage. Herbal medicines have been used for the treatment of diabetic patients since long and they are currently accepted as an alternative therapy for diabetic treatment. However in the indigenous Indian system of medicine good number of plants was mentioned for the cure of diabetes and some of them have been experimentally evaluated and active principle were isolated. WHO has also recommended the evaluation of the effectiveness of plants in conditions where there are no safe modern drugs. The ethnomedicinal information reports state that about 800 plants may possess anti-diabetic potential. Recently the medicinal values of various plants extracts have been studied by many scientists in the field of diabetic research. (Diabetes information hub) Diabetes is associated with long term damage such as malfunction of eyes, kidneys, nerves, heart and blood vessels.

Diabetes mellitus is a disease of all ages with a high percentage of prevalence in worldwide about 2.8 per cent in 2000 and expected to be 5.4 per cent in 2025. Allopathic drugs used for the treatment of diabetes were insulin and anti-diabetic drugs which are in almost conditions were expensive and cannot be reached easily specially in developing countries, marked by series side effects. Moreover, the management of diabetes is a global problem until now and successful treatment is not yet discovered. There are many synthetic medicines has been developed for patients. Medicinal plants used to treat diabetic conditions are of considerable interest and a number of plants have shown varying degrees of hypoglycemic and anti-hyperglycemic. The use of herbs as hypoglycemic is a major avenue in Indian perspectives particularly for treating diabetes, which require to be explored more effectively as there are so many literatures available on these aspects. Traditional plant medicines are used through-out the world for a range of diabetic presentations. Herbal drugs are prescribed widely because of their effectiveness, less side effects and relatively low cost therefore, investigation on such agents from traditional medicine plants has become more important. During the field survey, the list includes various types of preparations used to cure anti-diabetic. Herbal medicines prescribed by Paliyar tribal healers are either preparation based on single plant part or a combination of several plant parts. The dominant plant families used for treatment of diabetes were belonging to Asteraceae family. Ethnobotanical survey is most useful for scientists, research scholars and scientific companies for further studies on isolation and identification of active compounds that can be formulated into anti-diabetic drugs. Isolated drugs can use for diabetic patients.

Abrus precatorius L. (Kudumani) belongs to the family Fabaceae can be taken as mode of utilization as juice 2 teaspoons given orally twice a day till cure. *Acacia nilotica* (L.) Delile. (Karuvealam) belongs to the family Mimosaceae can be taken as Infusion of bark is drunk to treat. *Adhatoda vasica* Nees. (Adhatoda) belongs to the family Acanthaceae, leaf juice taken orally. *Aegle marmelos* L. correa (Villvam) Rutaceae leaves chewed and swallowed every morning help to reduced sugar of diabctcs paticnt. *Aerva lanata* (L.) Juss. Ex schult. (Poolai) belongs to the family Amaranthaceae pepper and leaf decoction is consumed regularly. *Allium cepa* L. (vengayam) belongs to the family Liliaceae juice is taken orally with honey or milk till cure. *Allium sativum* L. (Poondu) belongs to the family Liliaceae leaf and bulb taken orally to treat diabetes. *Aloe vera* L. (Kathalai) belongs to the family Liliaceae raw leaf bulb is consumed twice daily. *Andrographis paniculata* (Burm. F.) wall. Ex Nees. (Siriyanangai) belongs to the family Acanthaceae crush or powder with honey mixture eaten. *Annona squamosal* L. (Kaattuseetha) belongs to the family Annonaceae past taken orally with milk daily in the morning. *Aristolochia bracteolate* Lam. (Aduthinakodi) belongs to the family Aristolochiaceae Decoction is used for drinking. *Aristolochia indica* L. (Isuramuli) belongs to the family Aristolochiaceae ground seeds are mixed with black pepper and made into a paste and given three times a day for about 15 days to treat. *Asparagus racemosus* willd. (Thnervitan kezhanku) belongs to the family Liliaceae decoction of the tuber of and bark of *Azadirachta indica* is given twice a day. *Azadirachta indica* andr. Juss. (Veambu) belongs to the family Meliaceae green leaf mixed Curcuma aromatic juice given daily in the morning. *Boerhavia diffusa* L. (Mukuruttai) belongs to the family Nyctaginaceae two teaspoonful whole plant juice is given orally twice a day for 40 days. *Bombax ceiba* L. (Illavam) belongs to the family Bombacaceae 50g fresh roots are crushed then mixed with 500 ml water and boiled. One cup decoction is administered internally once a day to control the disease. *Caesalpinia pulcherrima* (L.) Sw. (Mayilkondrai) belongs to the family Caesalpiniaceae the powder is given with milk or honey. *Cajanus cajan* (Linn.) Millsp. (Kollu) belongs to the family Fabaceae cooked and taken along with food. *Cassia auriculata* L. (Aavaaramboo) belongs to the family Caesalpiniaceae one tea spoon powder of flower is given once in the morning for about 15 days or more Can reduce the diabatic level. *Cassia fistula* L. (Konrai) belongs to the family Caesalpinaceae powder of seeds is given once in the morning for about 15 days or more. *Cassia tora* L. (Nellatharai) belongs to the family Caesalpinaceae juice 10g mixed 400ml of cured water, is boiled and taken orally once a day for about 15 days. *Cantharanthus roseus* (L.) G. Don. (Nithyakalyani) belongs to the family Apocynaceae 10 gm flower is mixed with 100 ml hot water and given orally. *Centella asiatica* (L.) Urban. (Vallari) belongs to the family Apiaceae juice is taken in empty stomach. *Citrullus colocynthis* (L.) Schrad.

(kumapatti) belongs to the family Cucurbitaceae one seed taken raw eaten early in the morning for three weeks is good to cure the disease. *Coccinia grandis* (L.) J. Voigt. (kovaikai) belongs to the family Cucurbitaceae 20ml whole plant extract is given orally to treat. *Coccinia indica* Wt and Arn. (Malaikovai) belongs to the family cucurbitaceae juice of green fruit is given daily in the morning. *Curcuma longa Auct.* Non L. (Manjal) belongs to the famiy Zingiberaceae the powder and mixed with *Tinospora cordifolia* soaked in water for few hours and then filtered the filtrate is given to drink with milk. *Cynodon dactylon* Pers. (Aruku) belongs to the family Poaceae plant juice is mixed with *Leucas aspera* given in the morning. *Ficus benghalensis* L. (Aalam) belongs to the family Moraceae the powder is given to drink with milk. *Ficus racemosa* L. (Arasam) belongs to the family Moraceae dried fruits taken with warm water are helpful. *Gymnema sylvestre* (Retz.) R. Br.Ex Roemer and schultes. (Sarkkaraikolli) belongs to the family Asclepiadaceae leaves are eaten as raw directly or 5g dried leaf powder is mixed with water and given orally once in a day. *Helicteres isora* L. (valamburi) belongs to the family Sterculiaceae decoction can be given twice a day. *Hemidesmus indicus* (L.) Schult. (Nannar) belongs to the family Asclepiadaceae juice of bark is mixed with water. *Hibiscus rosasinensis* L. (Semparuthi) belongs to the family Malvaceae one tea spoon powder of flower mixed with pepper is given once in the morning for about 10 days or more. *Justicia adhatoda* L. (Aduthinapala) belongs to the family Acanthaceae plant juice mixed with pepper given in the morning and evening. *Lawsonia inermis* L. (Maruthani) belongs to the family Lythraceae leaf juice mixed with cow's milk taken once in a week. *Leucas aspera* Spreng. (Thumbai) belongs to the family Lamiaceae whole plant extract is taken in empty stomach. *Mangifera indica* L. (Maa) belongs to the family Anacardiaceae the infusion is made from boiled leaves and taken orally. *Melia azedarach* L. (Mlaiveambu) belongs to the family Meliaceae juice taken orally. *Mirabilis jalapa* L. (Anthimanatharai) belongs to the family Nyctaginaceae crushed tuber mixed with hot water can be taken after meals. *Momordica charantia* L. (Paagarkkaai) belong to the family Cucurbitaceae, leaf and fruit juice in early morning is used to treat. *Momordica dioica* Roxb. ex. Willd. (Naripagal) belongs to the family Curcurbitaceae tuberous root is ground in hot water and 50ml of the juice is taken orally once a day on an empty stomach for 10 days to treat. *Moringa oleifera* Lam. (kattumurungai) belongs to the family Moringaeceae boiled in water prepare the decoction taken as hot drink. *Murraya koenigii* (L.) Spr. (Kariveppilai) belongs to the family Rutaceae leaf extract 2-3 tea spoon early in the morning. *Nelumbo nucifera* Gaertn. (Thamarai) belongs to the family Nelumbonaceae made to juice and taken orally for about 15 days. *Ocimum sanctum* L. (Thullasi) belongs to the family Lamiaceae leaf powder is taken orally with honey. *Oxalis corniculata* L. (Pulichan) belongs to the family Oxalidaceae 100gm leaves are wrapped with banana leaf and after roasted in firewood is given with salt once daily at an interval of 3 days. *Phyllanthus emblica* L. (Nelli) belongs to the family Euphorbiaceae, one teaspoonful of juice obtained from squeezed leaves is taken in the morning on an empty stomach. Fruits are chewed and taken as much as possible, dried fruits are soaked in water and the water is taken orally with the fruit. *Piper betle* L. (Vettrilai) belongs to the family Piperaceae crushed and mixed with pepper powder boiled decoction empty stomach can reduce the diabatic level. *Psidium guajava* L. (Koya) belongs to the family Myrtaceae decoction is consumed regularly. *Pterocarpus marsupium* Roxb. (Seavankai) belongs to the family Fabaaceae bark decoction mixed with *curcuma longa* can drink daily. *Punica granatum* L. (Madhulai) belongs to the family Punicaceae powder about 1 teaspoonful is taken with water in the morning in empty stomach and also in the eveining before meals. *Ricinus communis* L. (Amanaku) belongs to the family Euphorbiaceae water extract of the plant is given twice daily, seeds powder is taken with milk. *Rubia cordifolia* L. (sevalaikodi) belongs to the family Rubiaceae powder is given with honey to treat diabetes. *Solanum nigrum* L. (Manattakkali) belongs to the family Solanaceae juice is taken orally in early morning. *Syzygium cumini* (L.) Skeels. (Naval) belongs to the family Myrtaceae powder about 1 teaspoonful is taken with water in the morning in empty stomach and also in the evening before meals. *Tamarindus indicus* L. (Puli) belongs to the family Caesalpiniaceae, boiled water, flower are mixed and then filter the decoction drink empty stomach. *Tephrosia purpura* Pers. (Kozhunchi) belongs to the family Fabaceae half cup of juice obtained from crushed stems is taken in the morning on an empty stomach. *Terminalia arjuna* (Roxb.) Wight and Arn. (Marutham) belongs to the family Combretaceae water extract of bark mixed with milk is given before sleep for 3 weeks. *Terminalia ahebula* Retz. (Kadukai) belongs to the family Combretaceae fresh fruits are also dried when available and when not available, dried fruits are soaked in water overnight followed by drinking the water in the morning. *Tinospora cordifolia* (Willd.) Hook. F and Thoms. (Kattukodi) belongs to the family Menispermaceae leaf decoction is given internally. *Tribulus terretris* L. (Nerungil) belongs to the family Zygophyllaceae dried fruit decoction is used for drinking. *Withania somnifer* Dunal (Amukkuran) belongs to the family Solanaceae, the juice of the leaves used for diabetes. *Zingiber officinalis* Roxb. (Iengi) belongs to the family Zingiberaceae, decoction is consumed regularly. *Zizyphus jujube* (L.) Gaertn. (Illanthai) belongs to the family Rhamnaceae powder is taken twice or thrice daily in empty stomach.

6. Indian Region - Store House of Traditional Knowledge

Indian region is one of the greatest emporia of ethno botanical wealth and a store house of traditional knowledge. Many adivasi groups (80 per cent of the population in villages) still live in isolation from the influences of the modern world and maintain close association with the ambient vegetation for all their needs. The time tested traditional knowledge perculates from generation to generation through oral folk lore. The profound traditional knowledge on resource use and their conservation by the tribals provide modern scientists with unparalleled opportunities for research and product development. Added to this, Indian region supports an enormous biodiversity of ancient lineage. These features have provided unique opportunities for study of ethno botany – a science dealing with relationship between human societies and plants.

A synthetic discipline of ethno botany are Botany (Taxonomy), Conservation, Forestry and Agriculture, Medicine Archaeology, Religion and Social customs, Geography, Anthropology. Sociology and Culture, Language and Linguistics, Economics Mythology, Ecology.

Methods of ethno-botanical study are the field work in tribal areas. Scrutiny of the herbarium specimens in large national and international herbaria. Scrutiny of ancient literature including the unpublished dairies and travelogues, analysis of different tribal folklore. Scrutiny of ancient sculptures on temples, forts and paintings. Scrutiny of plants and plant parts from the archaeological sites and excavations.

Study of fossilized plant materials and remains (paleo-ethno-botany, archaeo-ethno-botany). Locating knowledgeable tribal persons is not easy; some tribes (Andaman and Nicobar Islands) are even hostile to outsiders. One has to live among the tribals for, long duration, freely mix with tem and acquire their confidence by respecting their rituals and ceremonies before any worthwhile ethno-botanical data is gathered.

Harvard University had to spend almost 12 years among the tribals of North East amazon for publishing is valuable ethno-botanical accounts (Schultes, 1956, 1962, 1963). R.J. Rodins of Missouri botanical garden has botanized for almost 25 years among the *Kwanyama ovambos* (Rodins, 1985). Other works of this nature are Gunther, (1945) Ethno botany of western Washington; turner and bell (1971) on the ethno botany of coast saltish Indians of Vancouver Island.

Utilization pattern of wild plants by tribals of India (AICRPE REPORT. 1994) are Edible 3900, Medicinal 7500, Material and Cultural requirements 700, Fibre and Cordage 525, Fodder 400, Pesticides and pisicides 300, Gum resins and dye 300, Incence and Perfumes 100.

Hortus malabaricus: contains magnificent illustrations and descriptions of 742 plants with medicinal and other properties. This work is based on the traditional knowledge of renowed, hereditary physician of Malabar region, Itty Achuden and written by H A Van Rheede in Latin in 12 volumes (595 pages) in 1678-1693. Prof. K.S.Manilal spent over 38 years rendering *Hortus malabaricus* to English (2005) and the to Malayalam (2008), identifying botanical names of plants and other details. Through this stupendous task, Prof. Manilal has opened a vast treasure-house of ethnic medicinal plants.

An abode of life saving plants in Himalaya have the richness of Himalayan medicinal plants is well known through the ages. Medicinal virtues of Himalayan plants was highlighted even in the great Indian epic Ramayana – when Lord Hanuman was sent to Himalaya for collecting the life saving "Sanjeevani" plant for treating Lakshman, the brother of Lord Rama who was fatally ill in the war between lord Ram and Ravana at present day Sri Lanka. Even today, the alpine meadows in the Himalaya hold innumerable "Sanjeevani – like" plants having miraculous effects for various ailments, which need to be properly identified, scientifically evaluated and global level products developed for human welfare.

Rhodiola is the sanjeevani plant? The answer is Rhodiola sp., a cold desert plant inhabiting very difficult and high altitudes is recently reported as the sanjeevani plant by Leh based defence institute of High altitude research (DIHAR). Rhodiola (local name- solo (Ladakhi name) is a wonder plant that has immune modulatory, adapto genic (adopting to difficult envt. condition) and radio protecting abilities due to prescence of secondary metabolites and phytoactive compounds. The plant has also been found to have anti-depressant and appetizer properties. The herb can mitigate the effect of radiation used in bombs in biochemical war fares. The adapto genic qualities of the plant can help the soldiers in adjusting to the low pressure and low oxygen environment.

Some drugs from Himalayan medicinal plants are *Saussurea costus, Abies webbiana, Ephedra, Inula racemosa, Podophyllum hexandrum, Nardostachys jatamensi, Valeriana wallichii, Nardostachys jatamansi, Viola, Abies spectabilis, Berberis aristata, Picrorhiza kurrooa, Eulophia dubia, Swertia chirayita, Aconitum spp* and their products are Koflet (syrup) Kuftone (cap.) liv-10(syrup) Livosin (syrup) Mentat(syrup) Octin expectorant (syrup) Orthoherb(cap.) Pilex (tab.) Pedzer(tab.) Tredeptone(tab.) Pee-Eee-forte(tab.) Pyopil(liquid). The manufacturers are Himalayan drug co. Dharmani drug res, and trg. Instt., allens labs (Pvt.) Ltd., Himalayan drug co., myncil pharmaceuticals, Walter bushoell Ltd. Himalayan drug co. Lyovak laboratories swastika formulation (p) Ltd. Aimii pharm, pvt. Ltd., Swastik formulation (p) Ltd. Amulet pharmaceutical.

Ethno botanical usage of *Aegle marmelos* among different ethnic tribes for diarrhoea and dysentery are Tribes of Kotadoon (UP) (root and unripe fruit powder), Tribes of Mizoram (root bark and stem bark decoction), Tribe of salsette island (Bombay) (fruit), Tribes of eastern Rajasthan (fruit pulp), Tribes of Mount Abu (Rajasthan)(fruit). Tribes of Hardol (root bark decoction) tribes of Kamrup (Assam); Sanavar, Korales (MP) (leaf decoction), tribes of Bansi (Rajasthan) (root and unripe fruit powder), tribes of Kita (Bihar) (roasted fruit pulp powder), tribes of Prakasam (AP) (fruit juice), tribes of south Garhwal (fruit), Tharus of Kheri (UP) (fruit), tribe of Duduvally (Jammu) (fruit decoction), Tharus of Balrampur (UP) fruit pulp.

Same species used for jaundice by different ethnic Tribes in India is *Phyllanthus amarus* in such places are tribes of Magadh region of Bihar, Charaka samhita, Tribes of Shetrunjaya and pachechham hills, Gujarat, local people of Kerala, tribes of Phulbani dist. Of Orissa, Sugalis, Nakkala of Andhra Pradesh, Valaya, Paliyan, Pulayan of Tamil Nadu.

Cross-cultural ethno botany: Bio prospection at species level can be used in scanning of entire biodiversity for short listing of species for a particular aliment, by comparative ethnobotaical study of different tribes within a nation or among different nations is bio-prospection. Such short listed species can be subjected to scientific scrutiny and further bioprospection at molecular and chemical levels. All developing countries should develop a 'traditional knowledge database' not only for sharing royalities (if any products arise) but also for rewarding the ethnic tribes.

Traditional knowledge digital library (TKDL) in CSIR- TKDL has taken up significant strides since its inception in 2001 for developing database of Indian traditional knowledge systems. The TKDL database today has digitized 5.15 lakh formulations/practices from the Indian systems of medicine including Ayurveda, Siddha, Unnani, *etc.* However, a database of traditional knowledge of all adivasi tribes in India is lacking. Unless such a database is developed, cross-cultural ethno-botanical studies is not possible. This forms a great hindrance to cross-cultural ethno-botanical studies in India.

Ethno botany and bio prospecting: Ethno-botany/ethno-pharmacology contributes to the discovery of many important plant-derived drugs. Conversion of traditional knowledge into a commercialized product making use of modern science is also bio-prospecting or biodiversity prospecting. Bio-prospecting has been going on for centuries, but only within the past 20 years has this activity gained prominence due to technological advanced in pharmaceutical, biotechnological and agricultural sectors.

Traditional knowledge and Bio-prospection: Examples from Indian flora where traditional knowledge has played a key role in bio prospection. *Phyllanthus amarus* complex for hepatitis 'B' virus, *Trichopus zeylanicus* for invigoration, *Podophyllum hexandrum* for liver disorders, *Coptis teeta* for stomachic, debility, *Hydnocarpous kurzi* for leprosy, febrifuge, *Aconitum chasmanthum* for rheumatic and fever and a few others.

7. Role of Ethnobotany in Search of Newer Medicinal Plants and Drug Discovery

Numerous adivasi tribes, specially in northeast India have not been seriously studied with respect to the medicinal plants used by them for all their aliments. Ethno botanical field work followed by prioritization of leads through cross cultural studies would certainly throw light on the potential newer medicinal plants for certain deadly diseases. Data on mode of use, dosage, duration of use, *etc.* would be rewarding and can provide interesting leads/clues for new drug development. Even one global level effective drug (say for diabetes) can change the economy of a country. World over, Ca.74 per cent of the herbal drugs are derived from the follow up of traditional leads only.

In Ethno botany and drug discovery, though some of the tribal medicines have been already incorporated in the organized systems of medicine, much larger number of folk medicines yet remains endemic to tribal pockets. Scrutinize these rich ethnic natural resources for further chemical, Pharmacological and Clinical studies for development of potential herbal medicines.

Plant – derived modern drugs and traditional medicine in world over high percentage of useful plant- derived modern drugs are discovered as a result of scientific follow up of traditional medicine (119 plant based drugs: 74 per cent from traditional herbs (Farnsworth,1990). About 50 pharmaceutical drugs from ethno botanical leads (Cox, 1994). In contrast, other approaches like phytochemical screening, massive biological screening of randomly collected plants, and phytochemical examination of plants with the aim of identifying new chemical compounds have not proved satisfactory (Farnsworth *et al.*, 1985).

Ethno-botanically important anti-cancer species are *Terminalia arjuna, Androgarphis paniculata, Catharanthus roseus, Ochrosia elliptica, Podophyllum peltatum, Vaccinium stamineum, Morinda citrifolia, Semecarpus anacardium, Calotropis gigantea, Nothapodytes foetida, Butea monosperma, Mesua ferrea, Aloe barbadensis, Cassia auriculata, Jatropha curcas, Mimosa pudica, Cassia senna, Bauhinia variegate, Aaphanamyxis polystachya* and their common names of those species are Arjuna kalmegh, Vinca, Ochrosia, May apple, Deerberry, Indian mulberry, Bhilwa, Madar, Vasane mara, palash, iron wood, tree rose, Indian aloe, Tarwar, Danthi, Mint, Senna leaves, Oprchid tree, Rohituka. Their respective families are Combretaceae, Acanthaceae, Apocynaceae, Apocynaceae, Berberidaceae, Ericaceae, Rubiaceae, Anacardiaceae, Asclepiadaceae, Icacinaceae, Papilionaceae, Clusiaceae, Liliaceae, Caesalpinaceae, Euphorbiaceae, Mimosaceae, Caesalpiniaceae, Caesalpiniaceae, Meliaceae. The plant used parts are bark, dried leaves, whole plant, trunk bark, dried rhizome, fruit, fruit, fruit, whole plant, bark, bark, leavers, bark, leaves root, leaves, seed oil, whole plant, leaves, root, bark, leaves.

Cordyceps sinensis (Berk.) Sacc. (Clavicipitaceae) are known as "caterpillar fungus" this mushroom has been highly priced for its medicinal; properties such as antibiotic, increase of physical stamina, anti-cancer, anti-asthma, *etc.* A single mushroom costs around rs.200/- Found in Arunachal Pradesh and Sikkim in high Himalayas mountains. Local people find the caterpillar fungus by watching the locations where the wild yaks graze. So far, no cultivation techniques have been developed; this fungus can boost the economy of the region if systematic cultivation is taken up.

The plant *Panax assamicus* Banerjee from the family Araliaceae are commonly known as ginseng is found in the hills of Arunachal Pradesh, Manipur, Meghalaya and Nagaland. The rhizome is used as a tonic and vitalizer by the local herbal practitioners. One kg of rhizome is sold for Rs.700/- to 1,000/- in Manipur and Nagaland; also sold in the form of powder @ Rs.1, 000/- per Kg at Dimapur and Kohima in Nagaland. No cultivation, collected entirely from wild; has become very rare and endangered especially in Manipur and Nagaland.

Panax pseudo-ginseng – a species of northeast India has great potential; highly invigorating and energy giving plant. *Indian viagara. Trichopus zeylanicus* of Agasthyamalai, Western Ghats has great potential for product development. The fruits provide high energy and act as anti-fatigue.

8. Conclusion

Future responsibilities for ethnobiologists are to inventorize the traditionally used biological resources and development of tribe wise data bases. Conserve and revitalize the traditional cultures. Safeguard the sustainability of traditional biological resources and traditional knowledge against misuse or over use by "modern societies". Act as custodians of the traditional knowledge and on behalf of tribals decide and distribute the benefits that may accrue for their knowledge. Future responsibilities for ethnobiologists are to identify knowledgeable resource persons in each region; provide some subsidy for pursuing their unique profession. Cross cultural studies involving all developing nations and ethnically isolated tribes for prioritizing species for detailed investigations for drug development (this must involve taxonomists, ethno botanists, Phyto chemists, Pharmacologists and Biotechnologists). Regional training programmes for generating manpower in taxonomy and ethno botany.

9. References

1. Medeiros PSG, Alencar N, Vandebroek I, Pieroni A, Hanazki N, De Albuquerque UP: The Use of Medicinal Plants by Migrant People: Adaptation, Maintenance and Replacement. Evidence-Based Complementary and Alternative Medicine. 2011, in press
2. Vandebroek I, Balick MJ: Globalization and Loss of Plant Knowledge: Challenging the Paradigm. PLoS One. 2012, 7 (5): e37643-10.1371/journal.pone.0037643.
3. Bano N, Ahmed A, Tanveer M, *et al*. Pharmacological evaluation of Ocimum sanctum. J Bioequiv Availab. 2017; 9: 387–92.
4. Boktapa, N and Sharma, AK. 2010. Wild medicinal plants used by local communities of Manali, Himachal Pradesh, India. Ethnobot Leaf., 14: 259 – 267.
5. Jain, SK. 1991. Dictionary of Indian folk medicine and ethnobotany, New Delhi: Deep Publications.
6. Kamraj, VP. 2000. Herbal medicine. Curr Sci., 78: 35 – 39.
7. Nautiyal, S. 1981. Some medicinal plants of Garhwal hills. A traditional use. J Sci Res Plant and Medi., 2: 12 – 18.
8. Rai, L and Sharma, E. 1994. Medicinal plants of Sikkim Himalaya: status, usage and potential, Dehradun: Bishen Pal Singh Mahendra Pal Singh.
9. Rawat, GS and Garg, GP. 2005. Medicinal plants: trade and commerce opportunities with India. Indian For., 131: 275 – 287.
10. Samant, SS, Butola, JS and Lal, M. 2008. Agrotechniques of commercially viable medicinal plants in the Indian Himalayan Region, Mohal- Kullu: G.B. Pant Institute of Himalayan Environment and Development, Himachal-Unit.
11. Seth, MK. 2003. Trees and their economic importance. Bot Rev., 69: 321 – 376.
12. Jain SK 1989 Ethnobotany: an interdisciplinary science for holistic approach to man-plant relationships. In: SK Jain (ed) Methods and approaches in ethnobotany. Society of Ethnobotany, Lucknow, Uttar Pradesh (Proc II Train Course, Lucknow, 1989) p 9–12

13. Jain SK, Sinha BK, Gupta RC 1991 Notable plants in ethnomedicine of India. Deep Publications, New Delhi
14. Mehrotra BN 1986 Research in pharmacognosy on Indian medicinal plants. In: BN Dhawan (ed) Current research on medicinal plants in India. Indian National Science Academy, New Delhi, p 6–20
15. Mukerji B 1953 Indian pharmaceutical codex I–indigenous drugs. Council of Scientific and Industrial Research, New Delhi
16. Ayodhya, S., Kusum, S. and Saxena, A. (2010). Hypoglycaemic activity of different extracts of various herbal plants. International Journal of Research in Ayurveda and Pharmacy. 1: 212.
17. Bnouham, M., Ziyyat, A., Mekhfi, H., Tahri, A. and Legssyer, A. (2006). Medicinal plants with potential antidiabetic activity-A review of ten years of herbal medicine research (1990-2000). International Journal of Diabetes and Metabolism. 14 (1), 1.
18. Ezuruike, U. F. and Prieto, J. M. (2014). The use of plants in the traditional management of diabetes in Nigeria: Pharmacological and toxicological considerations. Journal of Ethnopharmacology. 155 (2), 857-924.

Chapter 6

Assessment of Medicinal Plants for the Treatment of Asthma in Pallivilai, Kanyakumari, Tamil Nadu

R. Subitha Shajini[1], *A. Nyla Azmi*[2] *and A. Anisha*[3]

[1]*Assistant Professor, Women's Christian College, Nagercoil, K.K. Dist., Tamil Nadu*
[2]*Research Scholar, Women's Christian College, Nagercoil, K.K. Dist. Tamil Nadu*
[3]*Women's Christian College, Nagercoil, K.K. Dist., Tamil Nadu*

ABSTRACT

Asthma is a common chronic disease that empower air to go into and out of the lungs occasionally limited, causing hacking, wheezing and shortness of breath. This study aimed to identifiy plants used for asthma by the local people of Pallivilai, Kanniyakumari District, TamilNadu and to establish the uses and local names of these plants. A field study was carried out over a period of approximately 4 months (December – March, 2023-2024). During this period, demographic characteristics of participants, local plants names, plant parts used and preparation methods of the plant were investigated and recorded. A total of 43 medicinal plants belonging to 23 families were identified in the region. The most common families were Lamiaceae, Asteraceae, Fabaceae and Apocynaceae. This study identified not only the plants collected for medical purposes by local people of Pallivilai, but also the local names of these plants. The results are very much encouraging and helpful in the treatment of asthma. The plants flora and the traditional knowledge of Pallivilai people is threatened by such factors as grazing, expansion of new agricultural lands, unsustainable picking of plants to generate income and lack of interest. Steps should be taken immediately to ensure the inclusion of relevant flora within conservation designations and to conserve this valuable knowledge from loss.

Keywords: *Asthma, Conservation, Families, Pallivilai and Valuable knowledge.*

1. Introduction

Plant resources have remained an integral part of human society throughout history. After fulfilling the primary needs like food and shelter, man has sought for a suitable remedy among plants for curing various diseases (WHO, 2002). Ten percent of all vascular plants are used as medicinal plants (Fonnerga, 2007), and there are estimated to be between 350,000 (Joppa *et al.,* 2011) and almost half a million (Pimm *et al.,* 2014) species of them. Since ancient times, plants have been used in medicine and are still used today (Grover *et al.,* 2002). Traditional medicine is defined as indigenous medicine that is used to maintain health and to prevent, diagnose, and treat physical and mental illnesses directly from allopathic medicine based on theories, beliefs, and experiences (WHO, 2012). Traditional medicine has been used for thousands of years with great contributions made by practitioners to human health, particularly as primary health care providers at the community level and has maintained its popularity world-wide (WHO, 2008). According to Sofowora about 60-85 per cent of the population in every country of the developing world has to rely on traditional medicine. (Sofowora, 1982)

Plants are a valuable source of bioactive compounds used directly or indirectly to treat human ailments. Ethno-medicine, the use of plants for disease treatment, has been practiced by tribal communities since time immemorial. Throughout history, civilizations have explored plant-based remedies for deadly diseases, including asthma. Asthma, a chronic disease affecting airways, causes symptoms like wheezing, coughing, chest tightness, and shortness of breath. It claims millions of lives worldwide, including in India. A 2005 WHO survey estimated that 300 million people suffered from asthma, with 255,000 deaths, 80 per cent of which occurred in low- and lower-middle-income countries (Braman, 2006). Inhaled allergens and pollutants are major asthma triggers. While treatment exists, synthetic drugs are often expensive. Traditional plant-based medicines have long been used as cost-effective alternatives for managing diseases, including asthma.

Some authors claim that transmitted knowledge is the origin of medicine and pharmacy. Even today, hundreds of higher plants are cultivated worldwide for medicinal and pharmaceutical substances (Kinghorn, 2020). The therapeutic properties of plants led to medicinal drugs made from beneficial plants (Jones *et al.,* 2006). Until the 18th century, many plants' therapeutic effects and treatment methods were known, but their active compounds were not (Faridi *et al.,* 2010).

The Renaissance, with advances in chemical analysis and tools like the microscope, enabled the isolation of active plant compounds (Reeds, 1976). Since then, these active principles have been synthesized in laboratories to produce medicines (Atanasov *et al.,* 2015). The use of medicines expanded, and today, medicinal plants are largely displaced in modern medicine (Gertsch, 2009). However, pharmaceutical industries still rely on plant-derived active principles as raw materials (Arceusz *et al.,* 2010). In underdeveloped regions, modern synthetic medicines remain inaccessible, so traditional plant-based medicine continues due to its low cost (Manzano *et al.,* 2021).

Medicinal plants have many industrial uses, from traditional medicines, herbal teas, and nutraceuticals to pharmaceuticals. They provide valuable foreign exchange for developing countries, supplying drugs like quinine and reserpine, galenicals, and intermediates like diosgenin. The pharmaceutical industry utilizes them for extracting active constituents, developing semi-synthetic drugs, and screening natural products for pharmacologically active compounds. The global market for plant-derived chemicals–pharmaceuticals, fragrances, flavors, and colors–exceeds billions annually. Developing medicinal plant bioindustries requires facilities for bioprocessing, extraction, purification, and marketing. Limited healthcare systems increase reliance on traditional medicine. Over 9,000 plants have known medicinal uses, with many still unexplored.

Due to changing user perceptions, commercialization, and global socio-economic transformation, indigenous knowledge on sustainable resource use has severely declined (Gadgil *et al.,* 1993) and must be documented before it is lost. This study assesses the ethnobotanical use of various plants, as reported in scientific literature, for treating asthma among tribal communities in India.

2. Methodology

2.1. Study Area

Tamil Nadu, the 11th largest state in India, covers 130,058 km^2, with 2,148 km^2 of forest area. About 3,305 km^2 (15 per cent) falls under protected areas, including 8 Wildlife Sanctuaries, 12 Bird Sanctuaries, 5 National Parks, 3 Biosphere Reserves, and 1 Tiger Reserve (Muthu *et al.,* 2006).

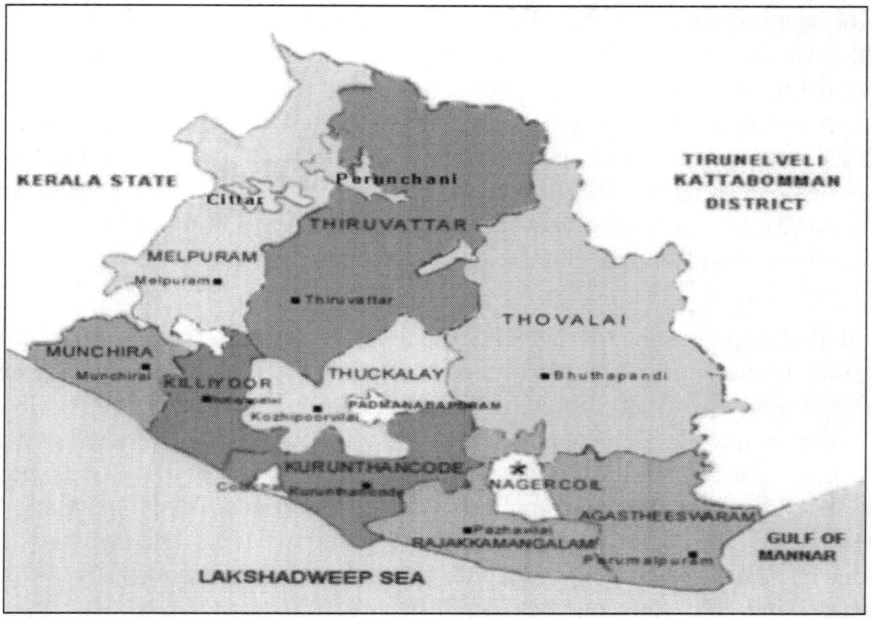

Figure 6.1: Kanyakumari District Map.

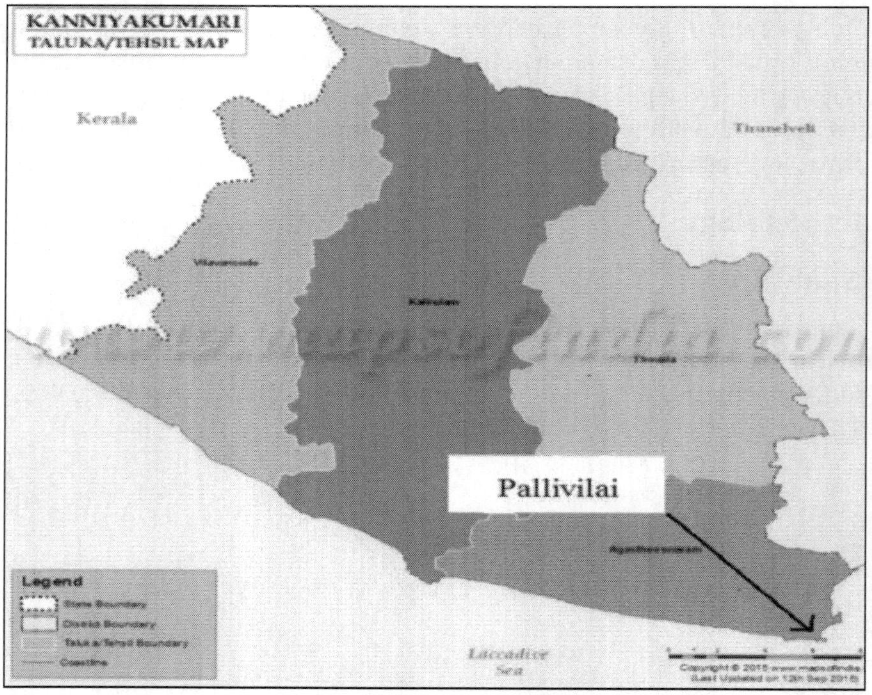

Figure 6.2: Study Area (Pallivilai).

Kanyakumari district, at the foothills of the Southern Western Ghats, spans 1,684 km^2 and marks the confluence of three water bodies and the Western and Eastern Coastal Plains (Sukumaran *et al.,* 2014). It consists of four taluks: Agastheeswaram, Thovalai, Vilavancode, and Kalkulam. This study focuses on Agastheeswaram taluk, specifically Pallivilai (18.7817°N, 83.4268°E), covering 7,392.35 hectares with a moderately hot and humid climate. Temperatures range from 26–27°C annually, peaking in January and February. Rainfall averages 1,313 mm, with maximum precipitation in October during the Northeast Monsoon.

Pallivilai has 7,851 inhabitants, with Tamil as the primary language. The population consists of 3,879 males and 3,952 females across 2,487 families. Rice (*Oryza sativa* L.) is the staple food, while some poorer residents consume tapioca (*Manihot esculenta* Crantz.). Traditional beverages like tea and chukka coffee (dried ginger and palm sugar) remain popular. The region's forests support diverse medicinal and aromatic plants, thriving in favourable climate conditions. The village follows a scattered settlement pattern, with agriculture and livestock farming as primary occupations. Traditional work includes rubber tapping, cane collection, and handicraft production. Seasonal harvesting of non-timber forest products such as honey, beeswax, medicinal plants, and gums is also common. Traditional medicine remains widespread in Pallivilai due to cultural acceptance, effectiveness, accessibility, and affordability compared to modern medicine. The region's diverse ecological, socio-cultural, and historical influences shape its medicinal practices, which have evolved through migration, foreign culture, and religion. Knowledge of herbal medicine is either secretly passed down orally or inherited through medico-spiritual manuscripts.

2.2. Data Analysis

Ethnobotanical surveys were conducted in Pallivilai over four months (December–March 2023–2024). Field trips were spread across seasons to maximize information and cross-check earlier findings with local informants. Informants were selected based on their folk knowledge distribution. A total of 20 individuals (9 males, 11 females), aged 48–75, were identified and asked to show plant species on-site. These informants possess extensive knowledge of local medicinal plants, practicing traditional medicine within their families. This knowledge, passed down orally for centuries, is now fading as younger generations show little interest. Data on ethno-medicinal plants, including local names, parts used (fresh/dried), preparation methods, dosage, and administration routes, were recorded. Collected specimens were identified and authenticated using valid references (Gamble and Fischer, 1957).

3. Result and Discussion

A survey identified 43 medicinal plant species from 23 families, highlighting their role as primary healthcare sources in the region. Elderly individuals possess extensive knowledge of these plants, commonly using them in the form

Table 6.1: Ethnobotanical Survey of Medicinal Plants Used for Treating Cough, Cold, and Related Ailments

Botanical Name	*Common Name*	*English Name*	*Family*	*Part Used*	*Uses*
Achillea millefolium L.	Achchilliyo	Yarrow	Asteraceae	Flower	Helps with cough.
Aerva lanata (L.) Juss. ex Schult.	Cerupulai	Mountain grass	Amaranthaceae	Whole plant	Leaves and root juices used for cough and throat infections.
Allamanda cathartica L.	Golden trumpet	Allamanda	Apocynaceae	Leaves, Bark	Bark used to prevent malaria and jaundice.
Allium sativum L.	Puntu	Garlic	Amaryllidaceae	Bulb	Crushed bulb used to prevent cold.
Althaea officinalis L.	Shemaitute	Marshmallow root	Malvaceae	Root	Tea from roots soothes sore throats and cough.
Amaranthus viridis L.	Kuppai kerai	Nettle	Amaranthaceae	Whole plant	Used to prevent jaundice, treat cough, and asthma.
Armoracia rusticana P. Gaertn., B. Mey. and Scherb.	Kutiraivali	Horseradish	Brassicaceae	Root	Roots help with cold, cough, and breaking fever by increasing respiration.
Azadirachta indica A. Juss.	Vepppai	Neem tree	Meliaceae	Whole plant	Neem juice used for dengue fever, cough, cold, and viral fever.
Catharanthus roseus (L.) G. Don	Nithya kalyani	Madagascar periwinkle	Apocynaceae	Root	Root decoction used to treat fever and cold.
Calotropis gigantea (L.) Dryand.	Erakku	Giant milkweed	Apocynaceae	Whole plant	Leaves used to treat asthma patients with severe cough.
Chamaemelum nobile (L.) All.	Chamomlie	Roman chamomile	Asteraceae	Flower	Chamomile tea helps treat colds.
Cissus quadrangularis L.	Pirandai	Creeper	Vitaceae	Stem	Used to prevent asthma and cold.
Cinnamomum verum J. Presl	Ilavankappattai	Cinnamon	Lauraceae	Bark	Warms the body and clears colds and flu.

Botanical Name	Common Name	English Name	Family	Part Used	Uses
Clitoria ternatea L.	Sangu puspham	Butterfly pea	Fabaceae	Leaves	Used to prevent cough, cold, and headaches.
Cymbopogon citratus (DC.) Stapf	Elumiccai pul	Lemongrass	Poaceae	Leaves	Lemongrass or oil used to treat fever and cold.
Eucalyptus globulus Labill.	Karpuramaram	Camphor tree	Myrtaceae	Leaf	Used to prevent cold, whooping cough, asthma, and malaria.
Euphorbia hirta L.	Sithirai palladai	Asthma plant	Euphorbiaceae	Leaves	Used to treat cough and cold.
Glycyrrhiza glabra L.	Atimaturam ver	Licorice root	Fabaceae	Root	Works as an expectorant and demulcent.
Hyssopus officinalis L.	Haicopas ahpicinails	Hyssop	Lamiaceae	Flower	Relieves cold, flu, and respiratory infections.
Matricaria chamomilla L.	Kemomil	Chamomile	Asteraceae	Flower	Used to prevent cold and fever.
Melissa officinalis L.	Elumiccai tailam	Lemon balm	Lamiaceae	Leaves	Tea helps with throat infections and cures colds.
Mentha spicata L.	Puthina	Spearmint	Lamiaceae	Leaves	Helps with colds and respiratory conditions.
Nepeta cataria L.	Peyimarutti	Catmint	Lamiaceae	Leaves	Used to prevent heavy fever and cold.
Ocimum basilicum L.	Tulasi	Basil	Lamiaceae	Leaves	Helps reduce fever and relieve symptoms of colds, coughs, and flu.
Origanum vulgare L.	Omam	Oregano	Lamiaceae	Leaves	Helps with cough, cold, flu, and minor fever.
Origanum majorana L.	Marikozhundhu	Marjoram	Lamiaceae	Leaves	Essential oil in bathwater helps ease cold and flu symptoms.
Piper betle L.	Vettilai	Betel	Piperaceae	Leaves	Boiled leaves with spices help relieve congestion and breathing issues.

Botanical Name	*Common Name*	*English Name*	*Family*	*Part Used*	*Uses*
Piper nigrum L.	Kurumilagu	Black pepper	Piperaceae	Fruit	Used to prevent cold and cough.
Plectranthus amboinicus (Lour.) Spreng.	Karpuravalli	Mexican mint	Lamiaceae	Leaves	Boiled leaves treat cold, asthma, and fever.
Psidium guajava L.	Guava	Lemon guava	Myrtaceae	Leaf	Prevents cough and throat infections.
Punica granatum L.	Mathulai	Pomegranate	Lythraceae	Skin of fruit	Used to prevent sore throats and cough.
Sambucus canadensis L.	Karupuu palam	Elderberry	Adoxaceae	Fruit	Used to prevent cold and cough.
Thymus vulgaris L.	Varatciyana taim	Thyme	Lamiaceae	Leaves	Thyme tea eases coughs, bronchial spasms, and congestion.
Trifolium pratense L.	Civappu klover	Red clover	Fabaceae	Flowers	Used in treating coughs and chest congestion.
Zingiber officinale Roscoe	Ingi	Ginger	Zingiberaceae	Root	Used for sore throats.

of decoctions, juices, powders, pastes, and whole-plant extracts. The Lamiaceae family had the highest representation (9 species), followed by Asteraceae and Fabaceae (4 species each), Apocynaceae (3 species), and several others with one or two species each.

The most frequently utilized plant parts were leaves, followed by flowers, whole plants, roots, fruits, bark, bulbs, stems, and rhizomes. Leaves were preferred due to their ease of preparation and high bioactive compound content. Herbs (60.46 per cent) were the dominant growth form, followed by trees, shrubs, and climbers.

Traditional preparations included decoctions (60.46 per cent), powders (37.20 per cent), oils, juices, extracts, and smoking. Remedies were administered in warm water or porridge, with dosage varying by age and ailment severity. Some treatments were applied topically as lotions or massages. Traditional healers emphasized that herbal remedies were safe and effective, with no adverse effects.

Notable species recorded include *Azadirachta indica, Catharanthus roseus, Plectranthus amboinicus, Phyllanthus niruri, Piper betle, Tamarindus indica, Calotropis gigantea, Amaranthus viridis, Cissus quadrangularis, Eucalyptus globulus, Ocimum basilicum, Piper nigrum, Cinnamomum verum, Allium sativum, Zingiber officinale, Cymbopogon citratus,* and *Mentha spicata.*

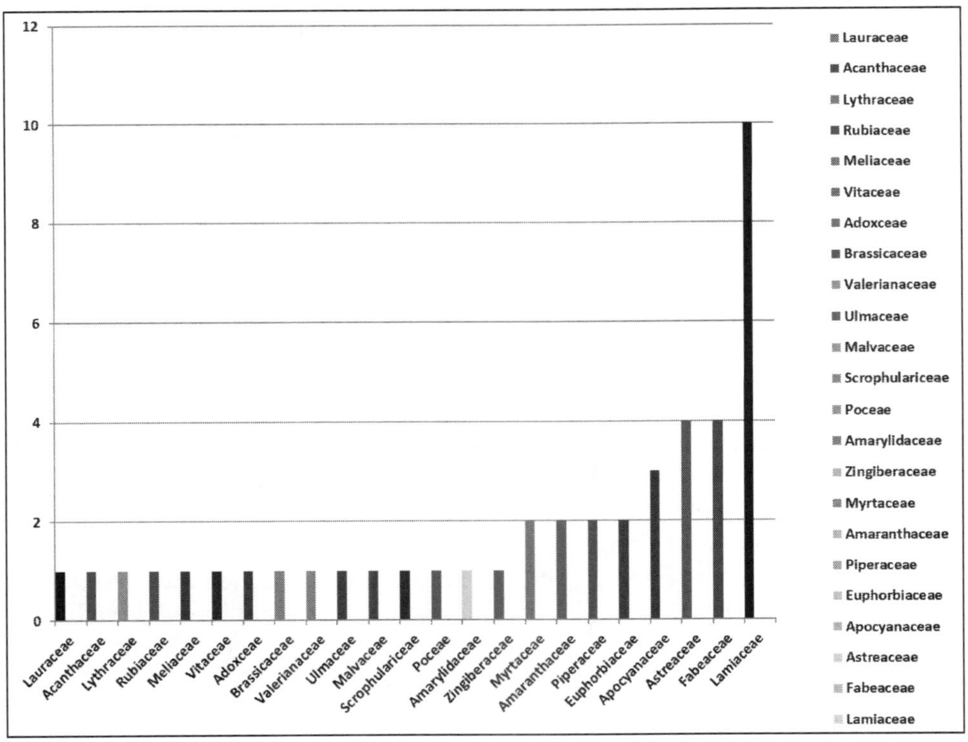

Figure 6.3: Family-wise Distribution of Ethnomedicinal Plant Recorded from Pallivilai.

This study highlights the importance of documenting indigenous knowledge for the conservation and sustainable use of medicinal plants. The reported species demonstrate pharmacological potential, particularly in treating respiratory conditions like asthma.

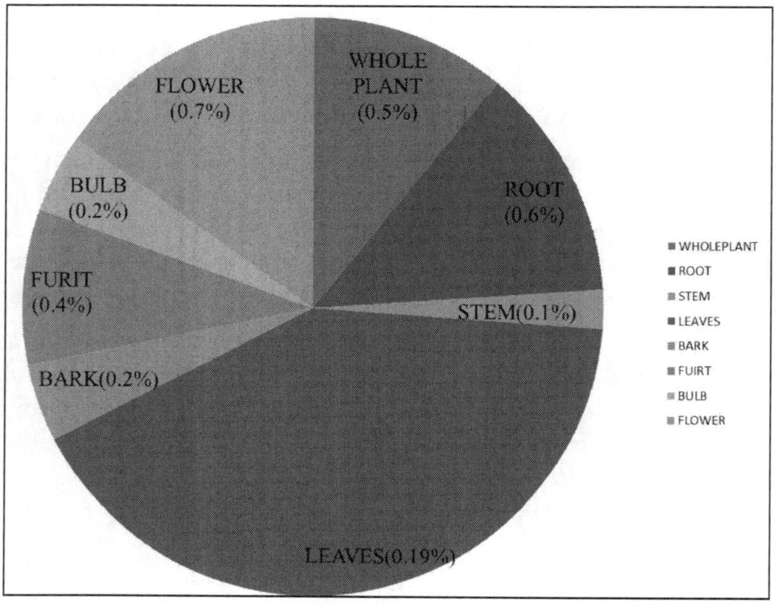

Figure 6.4: Growth Forms (Habit) of the Documented Medicinal Plant Species.

Fruits play a crucial role in local diets, providing essential nutrients and aiding in disease prevention. Monotherapy using a single plant species was the most common treatment, while polyherbal preparations combined two or more plants for enhanced efficacy, as supported by studies on synergistic effects in herbal medicine (Zonyane *et al.,* 2012; Bussman and Sharon, 2006). In such preparations, plants are dried, powdered, and stored separately, with dosage varying based on the ailment.

In Pallivilai, medicinal plants are widely accepted due to limited access to modern healthcare, poverty, and cultural beliefs. Rural communities rely on wild plant species for both human and veterinary medicine, recognizing their effectiveness and accessibility. Traditional knowledge, often passed down orally, plays a crucial role in healthcare, survival, and local problem-solving (Prakash, 2014). The medicinal plant usage in Pallivilai shows similarities with practices in Bhubaneswar, Odisha, where traditional healers also use plant-based remedies for asthma (Dash *et al.,* 2018). However, this knowledge is gradually fading.

This study documents the ethnobotanical practices of Pallivilai people in treating asthma, highlighting the potential of these plants for pharmaceutical applications. As modern medicine dominates healthcare, preserving and integrating traditional herbal knowledge may offer alternative treatments for asthma and other ailments.

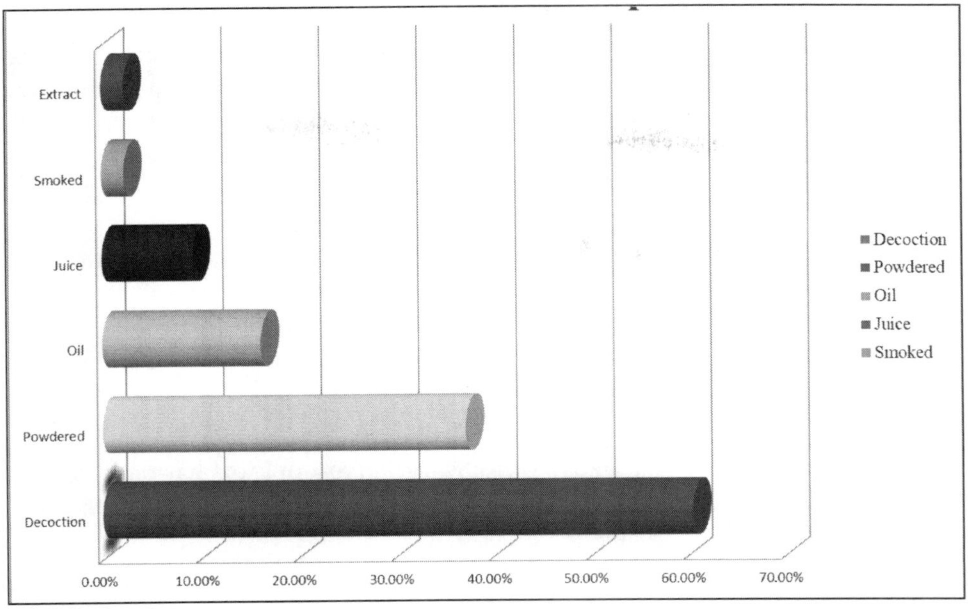

Figure 6.5: Method of Preparation of the Documented Medicinal Plant.

4. Conclusion

The people of Pallivilai have a deep connection with nature. A survey of the region revealed an abundance of medicinal plants used to treat a wide range of human ailments. Studies on traditional medicinal practices indicate that knowledge about these plants is primarily held by traditional healers, herbalists, and elderly individuals living in rural areas. Although this knowledge has been passed down orally from generation to generation, it has not yet been documented. Therefore, documenting this valuable information is essential for preserving it for future generations. Local residents, particularly the elderly, heavily rely on these traditional medicinal plants for their health. They believe these remedies are easily accessible, more affordable, and free from the side effects commonly associated with modern medicine. However, the study also highlights the over-exploitation of certain medicinal plant species by local residents, who may be unaware of their ecological significance.

Herbal medicine is one of the oldest forms of treatment used by humans. The study concludes that medicinal plants play a crucial role in the treatment of asthma in the Pallivilai region. While synthetic drugs are effective against asthma, they often have numerous side effects. In contrast, medicinal plants offer a safer and more cost-effective alternative.

The findings of this study suggest that the people of Pallivilai possess a rich traditional knowledge of medicinal plants. However, as urbanization and tourism expand in the area, younger generations are increasingly drawn toward market-driven economies, leading to a decline in traditional practices. Many local people now have to travel farther to collect medicinal plants, highlighting concerns about their depletion in the wild. Therefore, documenting traditional medicinal knowledge from this region not only recognizes its importance but also contributes to conservation efforts. Additionally, preserving this knowledge may provide valuable pharmacological insights for the advancement of modern medicine and the well-being of society.

5. References

1. Arceusz, A., Radecka, I., and Wesolowski, M. (2010). Identification of diversity in elements content in medicinal plants belonging to different plant families. *Food Chemistry, 120*(1), 52-58.
2. Atanasov, A. G., Waltenberger, B., Pferschy-Wenzig, E. M., Linder, T., Wawrosch, C., Uhrin, P., and Stuppner, H. (2015). Discovery and resupply of pharmacologically active plant-derived natural products: A review. *Biotechnology Advances, 33*(8), 1582-1614.
3. Braman, S. S. (2006). The global burden of asthma. *Chest, 130*(1), 4S-12S.
4. Bussmann, R. W., and Sharon, D. (2006). Traditional medicinal plant use in Northern Peru: Tracking two thousand years of healing culture. *Journal of Ethnobiology and Ethnomedicine, 2*, 47.
5. Dash, S. K., Sahoo, H. B., and Kar, D. M. (2018). Ethnobotanical study of medicinal plants used in the treatment of asthma in Odisha, India. *Journal of Ethnopharmacology, 222*, 1-12.
6. Fonnerga, A. (2007). The role of medicinal plants in human health: A global perspective. *Journal of Ethnopharmacology, 112*(3), 345-356.
7. Gadgil, M., Berkes, F., and Folke, C. (1993). Indigenous knowledge for biodiversity conservation. *Ambio, 22*(2-3), 151-156.
8. Gamble, J. S., and Fischer, C. E. C. (1957). *Flora of the Presidency of Madras* (Vol. 1–3). Adlard and Son Ltd.
9. Gertsch, J. (2009). How scientific is the science in ethnopharmacology? *Historical perspectives and epistemological problems. Journal of Ethnopharmacology, 122*(2), 177-183.
10. Grover, J. K., Yadav, S., and Vats, V. (2002). Medicinal plants of India with anti-diabetic potential. *Journal of Ethnopharmacology, 81*(1), 81-100.
11. Jones, W. P., Kinghorn, A. D., and Chin, Y. W. (2006). The role of pharmacognosy in modern medicine. *Current Drug Targets, 7*(3), 247-264.
12. Joppa, L. N., Roberts, D. L., Myers, N., and Pimm, S. L. (2011). Biodiversity hotspots house most undiscovered plant species. *Proceedings of the National Academy of Sciences, 108*(32), 13171-13176.

13. Kinghorn, A. D. (2020). The discovery of drugs from higher plants. *Advances in Pharmacology, 87*, 37-76.

14. Manzano, R. A., Figueroa, M. O., and Valenzuela, J. R. (2021). Traditional medicine as an alternative in regions with limited healthcare access. *Journal of Natural Products, 84*(7), 1962-1975.

15. Muthu, C., Ayyanar, M., Raja, N., and Ignacimuthu, S. (2006). Medicinal plants used by traditional healers in Kancheepuram District of Tamil Nadu, India. *Journal of Ethnobiology and Ethnomedicine, 2*, 43.

16. Pimm, S. L., Joppa, L. N., and Raven, P. H. (2014). The biodiversity of species and their rates of extinction, distribution, and protection. *Science, 344*(6187), 1246752.

17. Prakash, N. (2014). Traditional knowledge and practices: A sustainable approach to healthcare in rural communities. *International Journal of Herbal Medicine, 2*(4), 15-22.

18. Reeds, K. (1976). Renaissance humanism and botany. *Annals of Science, 33*(6), 519-542.

19. Sofowora, A. (1982). *Medicinal plants and traditional medicine in Africa*. John Wiley and Sons.

20. Sukumaran, S., Raj, V., and Jeeva, S. (2014). Plant diversity and vegetation analysis in Kanyakumari Wildlife Sanctuary, Western Ghats, India. *Journal of Threatened Taxa, 6*(8), 6050-6062.

21. World Health Organization. (2002). *Traditional medicine: Growing needs and potential*. WHO Policy Perspectives on Medicines, No. 2.

22. World Health Organization. (2008). *Traditional medicine fact sheet*.

23. World Health Organization. (2012). *WHO traditional medicine strategy 2014-2023*.

24. Zonyane, S., Van Vuuren, S. F., and Lall, N. (2012). Synergistic effects of herbal medicine: A review of its application in traditional medicine. *South African Journal of Botany, 78*, 212-217.

Chapter 7

Ethnobotanical Survey on Medicinal Plants Used to Treat Urinary Tract Infections in Kanyakumari District

R.L. Reshma[1] and V. Mary Kensa[2]

[1]Research Scholar (full time), Department of Botany, S.T. Hindu College Nagercoil, Affiliated to MS University, Tirunelveli– 629002, Tamil Nadu
e-mail: rejisha2002@gmail.com
Associate Professor, Department of Botany, S.T. Hindu College Nagercoil, Affiliated to MS University, Tirunelveli– 629002, Tamil Nadu
e-mail: surejkensa@gmail.com

ABSTRACT

This study is an attempt made to explore the indigenous knowledge of medicinal flora used to treat urinary tract infections in Kanyakumari district, Tamil Nadu, India. The information was collected by intensively interviewing the indigenous people, experienced aged rural folks, traditional healers and herbal drug sellers of Kanyakumari district. A total of29 medicinal plants belonging to 21 families were identified to be used as curative agents against urinary tract infections by dwellers of Kanyakumari district. The medicinal plants were arranged alphabetically based on its botanical name and other details such as local names in Tamil, family name, plant part used, method of preparation and route of administration were recorded. The indigenous knowledge obtained through this survey might aid the scientific community to investigate these plants for discovery of bioactive compounds that could serve as curative agents against urinary tract infections in future.

Keywords: *Kanyakumari, Indigenous knowledge, Medicinal flora, Urinary tract infections.*

1. Introduction

Throughout history, plant resources have remained an important aspect of human society. Following the fulfllment of necessities such as food and shelter, man has looked for appropriate medicine among plants to cure ailments. Medicinal plants have been utilized for many years to treat various ailments not only in rural areas but also increasingly in the urban areas of both developed and developing countries (Ralte *et al.*, 2024). Traditional system of herbal medicine utilizes cultural knowledge and practices for the maintenance of human health. This knowledge on medicinal plants is passed verbally from one generation to other and is still maintained by various indigenous groups (Ahmed*et al.*,2019).

Ethnomedicinal research has aided in the creation of both natural and synthetic medications. Interestingly, ethnobotanical knowledge has been used as an introduction for numerous successful drug screening studies. Likewise, ethnomedicine can be traced back to the origins of more than half of all pharmaceutical medications (Ralte *et al.*, 2024). Therefore, ethnomedicinal studies can be considered as an appropriate and reliable source of information on valuable medicinal plants that can be targeted to discover natural and synthetic drugs (Devi *et al.*, 2020).

Urinary tract infection (UTI) is one of the most prevalent bacterial infections in women and elderly individuals. This type of infection although can cause less severe life threatening infections but the patient experienced significant distress (Bazzaz*et al.*,2021). The studies showed that about 5 per cent of men and 40–50 per cent of women develop UTI in their lifetime and it accounts for more than 100,000 hospitalizations and spends about 1.6 billion dollars in medical expenses each year. UTI is caused by a variety of pathogens and uropathogenic *Escherichia coli* which accounts for approximately 75 per cent of the isolates and being the most common etiological agent followed by *Klebsiella pneumonia*. However, emergence of resistance to commonly used antibiotics in UTI has become a great challenge (Komal *et al.*, 2018).

Ethnobotanical knowledge on medicinal plants from different parts of India has been documented (Taid *et al.*, 2014). Ethnobotanical value of medicinal plants possessed by various tribals and rural communities were studied to a certain extent only in Tamil Nadu (Mahalik *et al.*, 2015). Scrutiny of these reports put forward the fact that the ethnobotanical knowledge on medicinal plants used to treat urinary tract infections in Tamil Nadu is inadequate, especially in Kanyakumari district is found to be sparse.

Hence, this study was conducted to explore the indigenous knowledge of medicinal flora used to treat urinary tract infections from folks residing in Kanyakumari district.

2. Materials and Method

2.1. Data Collection

The ethnobotanical investigation was carried out for a period of four months from September 2024 to December 2024.Information was learned from healers using semi-structured interviews with a questionnaire. The basic method followed was a guided field interview. The information was collected from the indigenous people, experienced aged rural folks, traditional healers and herbal drug sellers of Kanyakumari district. The information collected included formulations, ailments for which the formulations were used, and dosages dispensed. Information was also collected about any particular season for collecting plants, plant parts used and whether combination of plants were used to treat any particular ailment or if any single plant was used to treat multiple ailments. Afield data book was prepared to record the data on medicinal plants collected during the survey. A total of 80 informants were interviewed intensively and the data obtained regarding their knowledge on ethnomedicinal uses of plants along with their common names, morphology, parts used and dosage were recorded in the data sheet. The medicinal property of each plant was accepted as valid if at least five separate informants had a similar opinion. The morphology and medicinal uses of the plants were checked using the literature available

2.2. Study Area

The study area Kanyakumari district represents the southernmost tip of India. Geographically, the district lies between 77°15'and77°36'of the eastern longitudes and 8°03' and 8°35' of the northern latitudes. It is situated in the foot hills of southern Western Ghats and covers an areaofabout1684sq. km,whichis1.29percentofthetotalgeographical area of the state Tamil Nadu, India. The district is bound by Tirunelveli District on the North-East, Kerala on the North-West, Bay of Bengal in the South-East, Arabian Sea in the South-West and Indian Ocean in the South. The annual rainfall ranges from 103cm to 310cm and the altitude is 1829 meters above mean sea level. The district has a moderately humid climate. Topographically the district can be classified as mountainous region, middle region and coastal region. Due to diverse nature of landscape, the district is rich with fourteen types of forests that include tropical wet evergreen to thorn forests. The district is inhabited by Tamil and Malayalam. The location map of the study area, Kanyakumari district was illustrated in Figure 7.1.

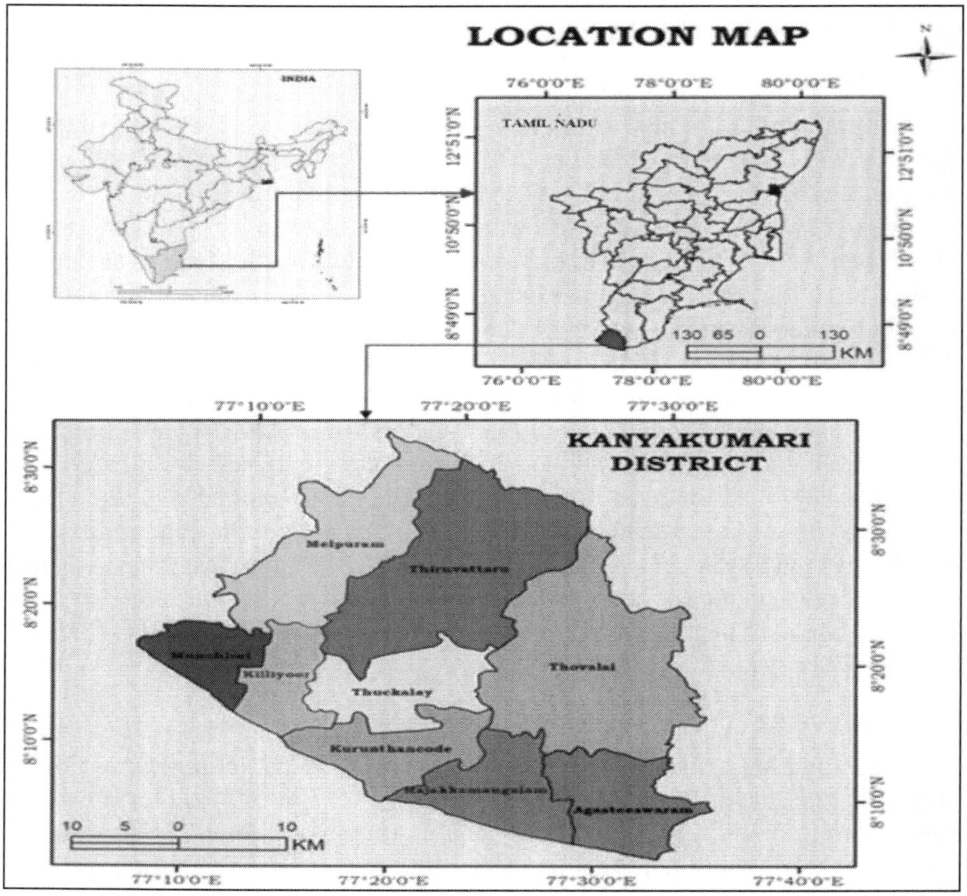

Figure 7.1: Location Map of the Study Area.

3. Result and Discussion

The traditional healers and indigenous elder people in Kanyakumari district possess rich knowledge on medicinal flora for curing various ailments. They believed this knowledge on herbal medicine as sacred enigmas and were reluctant to share. Only a few of them shared their traditional herbal knowledge after understanding the rationale of the study. The observation of this study revealed that traditional medicine plays a significant role in the life of rural folks residing in Kanyakumari district.

During the study, a total of 29 species of plants used against urinary tract infections by the traditional healers and rural folks residing in this district were identified. UTIs remedies included leucorrhea, frequent or infrequent urination, cloudy urination and burning sensations during urination. The plant species identified during the study were *Aerva lanata*, *Amaranthus viridis*, *Annona squamosa, Asparagus racemosus, Boerhavia diffusa, Borassus flabellifer, Canthium parviflorum, Centella asiatica, Clitoria ternata, Cocos nucifera, Coleus aromaticus, Coriandrum sativum, Cucumis sativus, Cuminum cyminum, Curculigo orchioides, Curcuma angustifolia, Cynodon dactylon, Hemidesmus indicus, Hordeum vulgare, Mangifera indica, Mimosa pudica, Moringa oleifera, Musa paradiosica, Phyllanthus emblica, Saraca asoca, Scoparia dulcis, Tribulus terrestris, Vetiveria zizanioides* and *Zingiber officinale.*

The medicinal plants identified through the survey were arranged in alphabetical order based on its botanical names along with family name, local names in Tamil, plant part used, method of preparation and route of administration in Table 7.1.

Table 7.1: Usage of Medicinal Plants against Urinary Tract Infections in Kanyakumari District

Sl. No.	*Botanical name with Family*	*Tamil Name*	*Plant Part Used, Mode of Preparation and Route of Administration*
1.	*Aerva lanata* (Amaranthaceae)	Sirupoolai	Leaves are soaked in water over night and the extract is filtered. About 100ml of the filtrate is consumed orally once in a day for 5 days.
2.	*Amaranthus viridis* (Amaranthaceae)	Kuppaikkirai	Water boiled with leaves and filtered. 100ml of the filtrate consumed 3 times daily for 3 days.
3.	*Annona squamosa* (Annonaceae)	Seetha pazham	50ml of decoction of leaves is consumed orally twice daily for7 days.
4.	*Asparagus racemosus* (Asparagaceae)	Shimaishadavari	5gm of root powder mixed in goat's milk is consumed orally once in a day for one month.
5.	*Boerhavia diffusa* (Nyctaginaceae)	Mukaratee-kirai	5 drops of honey added to 10ml of decoction of the whole plant and taken orally once in a day for 7 days.
6.	*Borassus flabellifer* (Arecaceae)	Nonkuppanai	Juice of ripe fruits mixed with sugar cane juice is consumed orally, till cure.
7.	*Canthium parvi-florum* (Rubiaceae)	Mullukaarai	Leaves cooked with rice flour and palm jaggery is taken internally for one week.
8.	*Centella asiatica* (Apiaceae)	Vallarai	50ml of leaf juice was consumed orally in emptyStomach for 7 days.
9.	*Clitoria ternatea* (Fabaceae)	Sanku puspam	5gms of dried root powder mixed in rice water and consumed orally one month.

Sl. No.	*Botanical name with Family*	*Tamil Name*	*Plant Part Used, Mode of Preparation and Route of Administration*
10.	*Cocos nucifera* (Arecaceae)	Tennai	Tender coconut added with cardamom, kept undisturbed overnight and drinks it in empty stomach for 3 days.
11.	*Coleus aromaticus* (Lamiaceae)	Karpoora valli	50ml of juice of leaves taken orally once in a day for one week.
12.	*Coriandrum sativum* (Apiaceae)	Kothamali	Water boiled with seeds and consumed as much as the patient could, till cure.
13.	*Cucumis sativus* (Cucurbitaceae)	Vellari	Grind ten seeds with little rock salt and consume the paste twice daily for 5 days.
14.	*Cuminum cyminum* (Apiaceae)	Jeerakam	Water boiled with seeds and consumed as much as the patient could, till cure.
15.	*Curculigo orchioides* (Amaryllidaceae)	Nilapanai	10gms of root powder mixed in cow's milk is consumed two times daily for 7 days.
16.	*Curcuma angustifolia* (Zingiberaceae)	Araukizhangu	5gms of root powder mixed with cow's milk is consumed two times daily for 7 days.
17.	*Cynodon dactylon* (Poaceae)	Aruvaumpullu	10ml of leaf juice is consumed orally in empty stomach for 5 days.
18.	*Hemidesmus indicus* (Asclepiadaceae)	Nannari	Root powder mixed with cow's milk is consumed orally twice daily for 7 days.
19.	*Hordeum vulgare* (Poaceae)	Barliarisi	Grains soaked in water overnight, boiled and drink as much as, till cure.
20.	*Mangifera indica* (Anacardiaceae)	Maa	Drink 500ml of well water after chewing seven leaves.
21.	*Mimosa pudica* (Fabaceae)	Thottalchinungi	50ml of decoction of leaves is consumed orally in empty stomach for 3 days.
22.	*Moringa oleifera* (Moringaceae)	Morigkai	Water boiled with leaves and consumed as much as, till cure.
23.	*Musa paradisiaca* (Musaceae)	Vaazha	20ml of stem juice consumed twice daily for three months.
24.	*Phyllanthus emblica* (Phyllanthaceae)	Nelli	Fruit juice mixed with honey is orally consumed twice daily, till cure.
25.	*Saraca asoca* (Caesalpinaceae)	Asogam	Seed powder is added to tender coconut and consumed three times daily.

Sl. No.	Botanical name with Family	Tamil Name	Plant Part Used, Mode of Preparation and Route of Administration
26.	*Scoparia dulcis* (Plantaginaceae)	Kallurukki	Fresh juice of leaves is consumed in empty stomach, till cure.
27.	*Tribulus terrestris* (Zygophyllaceae)	Nerunji	Juice prepared by mixing fruit powder with honey and milk, drink 100ml of this juice for one week.
28.	*Vetiveria zizanioides* (Poaceae)	Vattiver	Roots soaked in water, boiled and drink as much as, till cure.
29.	*Zingiber officinale* (Zingiberaceae)	Inji	2gms of root powder added to 500ml of water and consumed at night for 15 days.

Table 7.2: Medicinal Plants Collected from the Study Area

Sl.No.	Botanical name with Family	Medicinal Plant
1.	*Aerva lanata* (Amaranthaceae)	
2.	*Amaranthus viridis* (Amaranthaceae)	
3.	*Annona squamosa* (Annonaceae)	

Sl.No.	Botanical name with Family	Medicinal Plant
4.	*Asparagus racemosus* (Asparagaceae)	
5.	*Boerhavia diffusa* (Nyctaginaceae)	
6.	*Borassus flabellifer* (Arecaceae)	
7.	*Canthium parviflorum* (Rubiaceae)	
8.	*Centella asiatica* (Apiaceae)	

Sl.No.	*Botanical name with Family*	*Medicinal Plant*
9.	*Clitoria ternatea* (Fabaceae)	
10.	*Cocos nucifera* (Arecaceae)	
11.	*Coleus aromaticus* (Lamiaceae)	
12.	*Coriandrum sativum* (Apiaceae)	
13.	*Cucumis sativus* (Cucurbitaceae)	

Sl.No.	Botanical name with Family	Medicinal Plant
14.	*Cuminum cyminum* (Apiaceae)	
15.	*Curculigo orchioides* (Amaryllidaceae)	
16.	*Curcuma angustifolia* (Zingiberaceae)	
17.	*Cynodon dactylon* (Poaceae)	
18.	*Hemidesmus indicus* (Asclepiadaceae)	

Sl.No.	Botanical name with Family	Medicinal Plant
19.	*Hordeum vulgare* (Poaceae)	
20.	*Mangifera indica* (Anacardiaceae)	
21.	*Mimosa pudica* (Fabaceae)	
22.	*Moringa oleifera* (Moringaceae)	
23.	*Musa paradisiaca* (Musaceae)	

Sl.No.	Botanical name with Family	Medicinal Plant
24.	*Phyllanthus emblica* (Phyllanthaceae)	
25.	*Saraca asoca* (Caesalpinaceae)	
26.	*Scoparia dulcis* (Plantaginaceae)	
27.	*Tribulus terrestris* (Zygophyllaceae)	
28.	*Vetiveria zizanioides* (Poaceae)	

Sl.No.	*Botanical name with Family*	*Medicinal Plant*
29.	*Zingiber officinale* (Zingiberaceae)	

From the study, it can be observed that medicinal florae play a vital role in the lives of indigenous population of the district. Habit wise analysis of the data revealed the dominance of herbs, followed by shrubs, grasses, trees, climbers and creeper. Family wise analysis of thedatashowedthedominanceofPoaceaeandApiaceaefamilywith3 species, followed by families like Amaranthaceae, Arecaceae, Fabaceae and Zingiberaceae with 2 species and other families like Amaryllidaceae, Anacardiaceae, Annonaceae, Asclepiadaceae, Asparagaceae, Caesalpinaceae, Cucurbitaceae, Lamiaceae, Moringaceae, Musaceae, Nyctaginaceae, Phyllanthaceae, Plantaginaceae, Rubiaceae and Zygophyllaceae represented with single species (Table 7.3 and Figure 7.2).

Table 7.3: Distribution of Dominant Families

Sl.No.	*Family*	*No. of Species*
1.	Apiaceae	3
2.	Poaceae	3
3.	Amaranthaceae	2
4.	Arecaceae	2
5.	Zingiberaceae	2
6.	Fabaceae	2
7.	Annonaceae	1
8.	Asparagaceae	1
9.	Nyctaginaceae	1
10.	Rubiaceae	1
11.	Lamiaceae	1
12.	Cucurbitaceae	1
13.	Amaryllidaceae	1
14.	Asclepediaceae	1

Sl.No.	*Family*	*No. of Species*
15.	Anacardiaceae	1
16.	Moringaceae	1
17.	Musaceae	1
18.	Phyllanthaceae	1
19.	Caesalpinaceae	1
20.	Plantaginaceae	1
21.	Zygophyllaceae	1

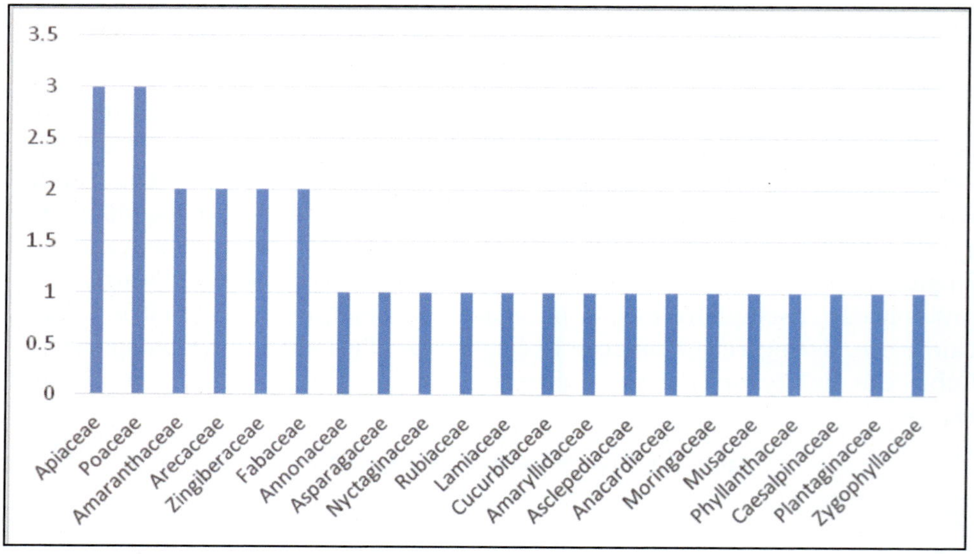

Figure 7.2: Distribution of Dominant Families

Apart from UTI, these plants were also used in treating respiratory system disorders, digestive system disorders, sex related problems, gynaecological problems, *etc*. The whole plant or parts like root, stem, bark, leaves, fruits and seeds were used to prepare decoction, powder or paste form of medicines. The plant parts were used either as fresh or as dried material for preparing herbal formulations. The plants were utilized either singly or in combination with other ingredients like milk, honey, jaggery, *etc*., for curing urinary tract infections. Most of the medicinal preparations were used as liquid forms during administration

Generally, the rural folks have strong belief on traditional healers and they in turn prescribed herbal medicines without any scientific validation for urinary tract infections. Based on experiences from patients, they validated the efficacy and success of drugs. Most of the people avoid visiting an allopathic doctor for treatment of urinary tract infections, because of their shyness to discuss about this infection with unknown physicians and being anxious about expenses of modern medicines.

4. Conclusion

From the study, it can be observed that numerous medicinal plants were used for treatment of urinary tract infections in Kanyakumari district. Based on severity and symptoms reported by victims, traditional healers prescribed the plants and its dosage. Hence it is necessary to investigate and document these plants to understand the basis of medicinal value of these plants by adopting modern scientific methods that may lead to the discovery of novel drugs with safety and efficacy. Moreover, this study will help to preserve the traditional knowledge on herbs, conserve the plants from extinction and provide income to the rural folks by cultivation of plants with medicinal value.

5. References

1. Ahmed, N, Anees M, Zhang, L 2019, An Appraisal of Ethnobotanical Investigation of Indigenous Flora from a High Temperature affected area in the Southern Punjab, Pakistan.PakistanJournal of Botany,Vol.51, no.4, pp.1493-1506.
2. Asha,S and Thirunavukkarasu, P 2013, Antiurolithiatic Activity of *Amaranthus viridis* on Ethylene Glycol Induced Male Rats,Inventi Rapid: Ethnopharmacology, no.4, pp1-6.
3. Bazzaz, BSF,Fork,SD, Ahmadi, R and Khameneh,B 2021, Deep insights into urinary tract infections and effective natural remedies,African Journal of Urology, Vol.27, no.6, pp.1-13.
4. Devi, PR, Kumar, SJ, Rejitha, S 2020, Existence and Survey of Medicinal Plants along the Neerody Coastal Line of Kanniyakumari District, International Journal of Advance Scientific Research and Engineering Trends, Vol.4, no.10,pp.10-20.
5. Dinnimath,BM, Jalalpure,SS and Patil,UK 2017, Antiurolithiatic activity of natural constituents isolated from Aerva lanata,The Journal of Ayurveda and Integrative Medicine, Vol.8, no.4, pp.226-232.
6. Dwivedi, PK, Salim, M 2016, An ethnobotanical survey on medicinal plants used in reproductivehealthrelated disordersin Dubri wild life sanctuary of Sidhi District, International Journal of Applied Research, Vol.2, no.11,pp.73-75.
7. Komal,S, Kazmi, SAJ, Khan, JA and Gilani, MM 2018, Antimicrobial activity of *Prunella Vulgaris* extracts against multi-drug resistant *Escherichia Coli* from patients of urinary tract infection, Pakistan Journal of Medical science, Vol.34, no.3, pp.616-620.
8. Heinrich,M 2000, Ethnobotany and its role in drug development, Phytotherapy Research, Vol.14, no.7,pp.479-488.
9. Hoareau, L, DaSilva, EJ 1999, Medicinal plants: a re-emerging health aid, Electronic Journal of Biotechnology, Vol.2, no.2, pp.56-70.
10. Jose,J and Devassykutty,D 2016, Evaluation of antibacterial activity of *Asparagus racemosus* in urinary tract infection,National Journal of Physiology pharmacy and Pharmacology, Vol.6, no.6, pp.596-598.

11. Mahalik, G, Sahoo, S, Satapathy, KB 2015, Ethnobotanical Survey of Plants used in Treatment of Urinary disorders in Dhenkanal district of Odisha, India, IOSR Journal of EnvironmentalScience, Toxicology and Food Technology, Vol.9,no.8, pp.58-63.

12. Ralte, L, Sailo, H and Singh, YT 2024, Ethnobotanical study of medicinal plants used by the indigenous community of the western region of Mizoram, India, Journal of Ethnobiology and Ethnomedicine, Vol.20, no.2, pp.1-24.

13. Samy, PR, Ignacimuthu, S 2000, Antibacterial activity of some folklore medicinal plant used by tribals in Western Ghats of India, Journal of Ethnopharmacology,Vol.,69,no.1,pp.63-71.

14. Taid, TC, Rajkhowa, RC, Kalita, JC 2014, A study on the medicinal plants used by the local traditional healers of Dhemaji district, Assam, India for curing reproductive health related disorders, Advances in Applied Science Research, Vol.5, no.1,pp.296-301.

Chapter 8

The Study Compares the Ethno-Medico-Botany of the *Lodha* People in West Bengal, India, with the *Apatani* and *Kanikkars* People in Other Regions of India

N. Ahamed Kabir[1] *and A. Doss*[2]

[1]*Research Scholar, PG and Research Department of Botany, V. O. Chidambaram College (Affiliated to Manonmaniam Sundaranar University, Tirunelveli), Thoothukudi, Tamil Nadu*

[2]*PG and Research Department of Botany, V. O. Chidambaram College, Thoothukudi, Tamil Nadu*

e-mail: kabir29092001@gmail.com

ABSTRACT

In the present study, the traditional medicinal plants used for treating various human ailments by the Lodha-populated in Paschim Medinipur, Jhargram, Bankura, Purulia, and Alipurduar villages in the South 24 Parganas of West Bengal, India, and the Apatani-populated in villages, namely Hong, Hari, Billa, Dutta, Hija, Mudang-Tage, and Michi Bamin in the Ziro valley of Lower Subansiri district in the central-western part of Arunachal Pradesh, India; and the Kanikkars populated in the Agasthiarmalai Biosphere Reserve, which is in the Western

Ghats of Kerala and Tamil Nadu, India; are scaned to know about the uniformity and variation in the ailments treated, part(s) used. The ethnomedicinal data of 45 plant species as having therapeutic properties against 81 human ailments under 21 ailment/health categories were collected from 3 research papers choosen for this review. Researchers in the area of ethno-medico-botany have not yet elaborated on this first comparative ethno-medico-botany of the Lodha populated in West Bengal with other tribal communities populated in Arunachal Pradesh, Tamil Nadu, and Kerala.

Keywords: *Indigenous knowledge, Ethnomedicinal knowledge, Traditional medicine, Tribal medicine, Ethnomedicine, Lodha, Apatani, Kanikkar*

1. Introduction

1.1. Indigenous Knowledge

Domestication is chiefly responsible for humanity's reliance on plants for survival, which goes back 10,000 years (Martin, 1995). Plants supply humans with food, insecticides, medicines, fuel, fodder, building materials, tools, and aesthetic and spiritual fulfillment. When people began to discover how to use plants, indigenous knowledge about plants emerged (Posey, 1999). Indigenous knowledge refers to the collection of information, rules, norms, abilities, and mental concepts held by native people in a certain place (Quanash, 1998). The immediate and intimate reliance of local people on natural resources resulted in the collection of indigenous knowledge that assisted people in adapting to their circumstances. In brief, ethnicity knowledge is unique in a specific culture or civilization and serves as the foundation for agriculture, healthcare, food preparation, education, environmental protection, and other diverse activities (Thomas, 1995). The complex knowledge, beliefs, and practices commonly referred to as indigenous knowledge evolve and change across time and space. As a result, such knowledge incorporates time-tested expertise gained via human contact with their surroundings (Alcorn, 1984). As a result, it is the culmination of many generations' long-term experiences, meticulous observations, and trial-and-error experimentation (Martin, 1995).

1.2. Ethnomedicinal Knowledge

Traditional medicine is one of the most extensively used indigenous knowledge systems. Such information, known as ethnomedicinal knowledge, entails traditional diagnosis, the collecting of raw materials, the production of medicines, and their prescription to patients (Fransworth, 1994a). Traditional knowledge about cures is passed down orally from generation to generation in many places, with tremendous secrecy (Jansen, 1981). Such secretive and crude transfers make indigenous knowledge or ethnomedicinal knowledge vulnerable to distortion and in most cases, some of the lore is lost at each point of transfer (Getahun, 1976), highlighting the need for systematic documentation of such valuable knowledge through ethnobotanical research.

1.3. Traditional Medicine

From time immemorial, man has had an endless yearning for an excellent and healthy life, which has led to his curiosity to investigate all areas of his surroundings via trial and error (Dalziel, 1961). That gave rise to traditional medicine, which was the only means to save lives before the arrival of modern medicine, as the first people employed various herbs to heal disease (Ajiwe *et al.,* 2008). Traditional medicine is defined by the World Health Organization (WHO) as "the total of knowledge, skills, and practices based on theories, beliefs, and experiences indigenous to different cultures, whether explicable or not, used in the maintenance of health as well as the prevention, diagnosis, improvement, or treatment of physical and mental illness." Nevertheless, the WHO warns that "inappropriate use of traditional medicines or practices might have detrimental or harmful effects" and that "additional study is needed to verify the efficacy and safety" of some of the traditional medicine systems' practices and medicinal plants (WHO, 2008). Traditional medicine that has been adopted by other people (beyond its indigenous culture) is generally labeled complementary or alternative medicine (CAM) (Gurib-Fakim, 2006). Traditional medical systems utilizing plants as a form of healing may be dated back to the Middle Paleolithic period, some 60,000 years ago, according to fossil findings (Solecki and Shanidar, 1975). Plants' traditional use is unavoidable in supplying folk remedies for the healthcare system, as well as a source of sustenance for the low-income class and rural areas. The indigenous system of folk remedies based on the usage of plants by local populations has been practiced for millennia and is passed down from generation to generation (Ugulu *et al.,* 2009).

1.4. Tribal Medicine/Traditional Medicine

Tribal medicine, also known as traditional medicine, is essential in the primary healthcare of the tribal and rural populations (Rajiv, 1998; Patil, 2008). Traditional medicine, which is practiced all over the world, has been acknowledged by the World Health Organization (WHO) as a crucial component of basic healthcare (Bannerman, 1982). According to the World Health Organization, traditional medicine is essential for the survival of 80 per cent of the world's population for primary healthcare (Hiremath and Taranath, 2013). Traditional treatments are becoming increasingly popular as a low-cost alternative to or supplement to mainstream therapy (Tang and Haliwel, 2010). Traditional medicinal traditions are now an essential component of supplementary or alternative medicine. Even though their efficacy and mechanisms of action have not been scientifically validated in the majority of cases, these basic pharmaceutical formulations frequently mediate favorable reactions due to their active chemical ingredients (Park and Pezzutto, 2002).

Traditional medicine is extensively practiced and accounts for around 40 per cent of all healthcare provided (WHO, 2002-2005). Traditional medicine is widely used, and plants continue to be a rich source of unique biological active chemicals with diverse activities such as anti-inflammatory, anti-cancer, anti-viral, anti-bacterial, and cardioprotective properties (Yan *et al.,* 2002). Around 85 per cent of the traditional medicines are acquired from plants (Fransworth, 1988), and over 80 per cent of the population in underdeveloped nations still uses plant-based traditional medicine for their healthcare (Silva, 1997). That is especially common in tribal people, whose livelihoods are usually based on woods, which have also helped to build their economical and cultural lives (Shroff, 1997; Mukul *et al.*, 2007). According to Kibebew (2001), traditional medicine is used by 75-90 per cent of the world's rural population (excluding Western nations). Not only are individuals unable to acquire pricey modern pharmaceuticals due to poverty, but the traditional methods are also more culturally acceptable and address psychological requirements in ways that modern medicine does not.

Traditional plant knowledge has become a recognized tool in the hunt for novel medication sources (Bruce and Meeus, 2002). Traditional medicinal plant knowledge has always influenced the hunt for novel treatments. Despite the introduction of current high throughput drug discovery and screening procedures, old knowledge systems have provided leads to the identification of useful medications (Buenz *et al.*, 2004).

1.5. Ethnobotany

Ethnobotany is a multidisciplinary field that studies the interactions between plants and humans in the context of their culture. It is difficult to pinpoint when the word ethnobotany entered modern science. It may, however, be traced back to the period when people began consciously interacting with plants and animals. Ethnobotanical work appears to have begun with (Christopher Columbus in 1492) when he brought tobacco, maize, spices, and other useful plants to Europe from Cuba (Cotton, 1996) and other immigrants from the new world documented food, medicine, and other useful plants of the Aztec, Maya, and Inca peoples (Cotton, 1996; Martin, 1995).

In 1895, Johann Harshberger coined the word ethnobotany for the first time (Balick, 1996). Yet, depending on the interests of the employees participating in the study, this phrase has been given many meanings. Harshberger (1896) defined ethnobotany as the study of plant utilization by indigenous people (Cotton, 1996). Yet, the field's history began far earlier. On the Indian subcontinent, ethnobotanical knowledge is a very old field. The ancient Indians should be credited for cultivating what is now known as ethnobotany (Kirtiker and Basu, 1993). Several plants' medicinal powers were mentioned in the Rigveda, Atharvaveda, Upanishads, Mahabharata, and Puranas, which were written between 4500 and 1600 BC (Jain, 1994). It was followed by Atharvaveda, which detailed the magical-religious features and therapeutic plant usage. Subsequently, the eight divisions of Ayurveda, recorded between 2500 and 900 BC, served as the true foundation of traditional medicine. Chikitsa Sthanam of Sushruta Samhita, written between 600 BC and 300 AD, has detailed details of medicinal herbs (Bishagratna 1907-1916; Jain 1996; Jain and Mudgal, 1999). People who study the relationships between humans and plants in any field (for example, medicine, food, and forestry) and at any level (local, regional, and so on) have recognized in recent years that the approach to ethnobotanical studies has undergone conceptual and methodological changes since the coining of the term Ethnobotany by Harshberger in 1896 (Carrió and Vallès, 2012).

Ethnobotany is the study of the link between plants and humans: "ethno" means "people study" and "botany" means "study of plants." Ethnobotany is a subfield of ethnobiology. Ethnobotany is the study of the intricate links between plants and human civilizations. Ethnobotany is the study of how plants are used, handled, and perceived in human civilizations. Plants used for food, medicine, divination, cosmetics, dyeing, textiles, construction, tools, money, clothing, rituals, social life, and music are examples of ethnobotany (Staff, 2007).

1.6. Importance of Ethnobotany

Ethnobotany is critical for understanding the dynamic interactions between biological variety, and social and cultural systems (Schultes, 1962; Balick, 1996; Hussain *et al.,* 2008; Mahmood *et al.,* 2011). It provides value to people in terms of traditional knowledge and understanding of their cultures, as well as practical plant usage. Ethnobotany also works with traditional medicine to respect knowledge and community (Cotton, 1996). Ethnobotanical research serves as a link between botany and tribal knowledge about plant therapeutic effects. Ethnobotany records traditional plant knowledge and may be used to involve policymakers and development planners in developing appropriate strategies for the conservation of cultures and cultural knowledge associated with their beneficial plants (Schultes and Reis, 1995). Ethnobotanists study the intricate links that exist between cultures and (plant uses), with an emphasis on how plants are utilized, maintained, and perceived among human communities. That encompasses food, clothes, currency, ritual, medicine, dye, construction, cosmetics, and a variety of other applications (Acharya and Anshu, 2008). According to Martin (1995), ethnobotanical research requires the participation of specialists from various disciplines, such as plant taxonomists, plant ecologists, anthropologists, linguists, economic botanists, pharmacologists, and others, to obtain more detailed and reliable information about plants and plant use. Ethnobotany uses transdisciplinary and multidisciplinary methodologies to collect and document indigenous botanical knowledge, cultural practice, usage, and management of botanical resources.

1.7. Tribal Community

A tribal community is a sociopolitical group of males who share a shared dialect, culture, and territory while also having a common language and artificial or real familial ties (Risley 1908). D. N. Majumdar claims that the tribal community is made up of some families that have come together to establish their distinct group, who all reside in the same region and speak the same language. They all follow the same procedures for marriage, work, and religion (Majumdar and Madan 1980). According to Pidington, a tribal community is a distinct group of people that speak the same language, live in the same area, and share the same social and cultural practices (Baskey, 1987).

1.8. Origin of the Term Tribe

Anthropologists define a "tribe" as a collection of people that typically reside in a certain area, speak the same dialect, and preserve the integrity of their organisation as well as cultural resemblance (http://inet.vidyasagar.ac.in:8080/jspui/bitstream/123456789/5449/17/17_Chapter per cent 206.pdf). A tribe is defined as a "group of families possessing a similar name, speaking the same dialect, inhabiting or purporting to occupy a common region and is not generally endogamous though it may have been so historically" (Nithya, 2014). The word "Tribe" comes from the Latin word "*Tribus*," which refers to a certain type of social and political organisation present in all of these cultures. In the beginning, it meant three divisions to the early Romans. Later, it came to refer to the "poor" or "masses." In the sixteenth century, the word "Tribe" first emerged in the English language. It signified that a group of people shared an ancestor (Biswal 1985; Parvathamma 1984). They go by a variety of names in India, including "*Adivasi*" (original settlers), "*Girijan*" (hill people), "*Vanyajati*" (forest caste), "*Adimjati*" (basic caste), "*Jati*" (folk groups), and "*Anusuchit Jan Jati*" (Scheduled Tribes) (Panda 2006). India has the second-largest tribal population in the world after Africa (Elwin, 1955).

1.9. Tribal People in India

According to anthropological literature, the word tribe has been used for groups of human beings whose domicile is in distant locations such as hills, forests, sea coastlines, and islands; and whose way of life is substantially different from that of modern civilized men. The forest plays a major role in tribal culture and economy. From birth to death, the tribal way of life is highly influenced by the forest (Nithya, 2014). Officials and anthropologists in British India used the phrase to refer to a social group that was being counted. The Government of India Act, of 1935 referred to them as "Backward Tribes." The Indian Constitution designated them as "Scheduled Tribes" and included them in a schedule, but it didn't define what a "Scheduled Tribe" was (Biswal 1985; Parvathamma 1984).

India is also known for having the world's second-biggest tribal (Adivasi) population. Paradoxically, the poorest people in India live in locations with the most natural resources. According to the 2011 census, tribals make up 8.2 per cent of the overall population. The Indian Constitution refers to indigenous people as "Scheduled Tribes." India recognizes 461 ethnic groups as Scheduled Tribes, who are considered India's original peoples. The Scheduled Tribes are generally referred to as Adivasis in mainland India, which means "original peoples." According to the Anthropological Survey, India today has a total of 4,635 groups. There are 732 "tribal" or Indigenous communities within this total. The seven states of northeast India, as well as the so-called "central tribal belt" running from Rajasthan to West Bengal, have the highest concentrations of indigenous peoples. Madhya Pradesh, Chhattisgarh, Maharashtra, Odisha, Jharkhand, and Gujarat are home to more than half of the Scheduled Tribes population. Around 700 Scheduled Tribes have been notified under Article 342 of the Indian Constitution, and they are dispersed across the country's states and union territories. Numerous tribes can be found in more than one state. The state of Orissa has the most Scheduled Tribes (62 in total) (Nithya, 2014). The Indian subcontinent is home to around 53.8 million tribal people living in 5000 tribal community forest-dominated villages, accounting for 15 per cent of the entire geographical area of Indian landmasses and representing one of the richest emporia of ethnobotanical resources (Sajem and Gosai, 2006).

Practically every sector of the Indian population employs plants as medicine and around 7500 plant species are used by various ethnic tribes. Tribal people, in particular, gather and conserve locally accessible wild and cultivated plant species and use herbal medicine to treat a wide range of ailments and problems. India, with its massively diverse ethnic groupings and abundant biological resources, is one of the major emporia of ethnobotanical riches (Kala, 2005). Most ethnobotanical research has focused on tribal and rural populations to document their knowledge and usage of plants and to hunt for new sources of herbal medications, food plants, and other plant qualities of use to man (Jain and Mudgal, 1999).

1.10. Ethnomedicine

Tribals have a strong conviction in their local folklore medicine and rely only on their herbal treatment (Sajem and Gosai, 2006). Numerous tribal tribes have used various plant or animal sources for medical formulations, which are known as ethnomedicines (Pushpangadan and Atal, 1984). Ethnomedicine can be described roughly as the use of plants as remedies by humans, however, these applications could be referred to more precisely as ethnobotanical medicine (Fransworth, 1994b). Ethnomedicine is one of the medical systems that are widely used to treat ailments among our country's tribal and aboriginal populations (Singh *et al.,* 2013). Ethnomedicine refers to plants, animal products, and minerals utilized by tribal tribes in a certain location or nation for medical uses different from those listed in the respective cultures' classical traditions. Ethnomedical information/ data are critical in the development of new scientifically verified and standardized medications, both herbal and contemporary (Savnur, 1993).

There is a growing effort in poor countries to incorporate traditional remedies, particularly herbal preparations, into local healthcare systems, and many contemporary researchers are interested today in exploring the enormous potential of ethnobotanical knowledge for treating many ailments (Dutta and Dutta, 2005; Kala, 2005; Jain *et al.,* 2010; Jeyaprakash, 2011). Unfortunately, ethnomedicinal plants are endangered owing to deforestation, overgrazing, and careless use. As a result, it highlights the critical necessity for their preservation. The preservation of traditional knowledge requires the conservation of biological resources as well as their appropriate usage (UNU-IAS, 2013). Ethnobotanists may play a decisive role in preserving ethnobotanical knowledge as part of live cultural and ecological systems, fostering pride in local cultural knowledge and practice, and maintaining the vital linkages between communities and the environment (Martin, 1995).

2. Materials and Methods

In the current study, the ethnomedicine data of 250 medicinal plants gathered from the *Lodha* populated settlements such as Paschim Medinipur, Jhargram, Bankura, Purulia, Alipurduar villages and South 24 Parganas of West Bengal, India documented by Chaudhury *et al.,* (2018) were compared with the ethnomedicine data gathered from the *Apatani* populated in seven villages, namely Hong, Hari, Billa, Dutta, Hija, Mudang-Tage, and Michi Bamin in Ziro valley of Lower Subansiri district in the central western part of Arunachal Pradesh, India; and the *Kanikkar* populated in Agasthiarmalai Biosphere Reserve, the Western Ghats, Tamil Nadu, India documented by Kala (2005) and Rani *et al.,* (2011) to know about the uniformity and variation in ailments treated and part(s) used that is yet to be made available to the modern world is collected by running multiple searches on different online databases like Google Scholar, ResearchGate, and ScienceDirect using relevant search terms.

2.1. *Lodha*

The *Lodhas* are a primitive tribe in India that mostly inhabits the states of West Bengal and Odisha. They primarily reside in West Bengal's Jhargram, Paschim Medinipur, Bankura, and Purulia districts. They primarily reside in the Mayurbhanj and Keonjhar districts of Odisha. They speak a damaged Mundari dialect mixed with Bengali, Oriya, and Hindi and appear to come from the Mundari ethnic group. Although they have been absorbed into the Hindu fold, they nonetheless retain unique characteristics. A slice of meat named after their ancestor is referred to as a "*Lodhas*." The tribal inhabitants of Jungle Mahal, who were once reliant on the forests, were brutally exploited by the British during the early British era, who then rose against them. As a result of being forced from their natural habitats and turning to criminal activities, the British Administrators labelled them as a criminal tribe. They have faith in Jarasandh's lineage from the Mahabharata. Nayek, Paramanik, Mallik, Digar, Sardar, Bhukta, Kotal, Dandapat, Bhuinya, and Ari are their titles. They were displaced from Jungle Mahal and mostly relied on foraging for roots, shoots, fruits, leaves, honey, *etc.* to survive. They perform agricultural and day labour in other people's fields (http://inet.vidyasagar.ac.in:8080/jspui/bitstream/123456789/5449/17/17_Chapter per cent 206.pdf).

The Savar an acknowledged primitive tribal group of Orissa is claimed as the ancestor of the *Lodhas*. The *Lodhas* was originally classified as a criminal in British India because of its reputation for criminal activity. According to the 2001 census, 7458 *Lodhas* people are living in Orissa (3650 men and 3808 women). The main hamlets are in the Community Development Blocks of Suliapada and Morada. Baripada, Badasahi, Khunta, Udala, and Kaptipada. Outside of Orissa, West Bengal and Jharkhand are also the states where the *Lodhas* reside. In general, their towns are quite close to woods that have a sufficient supply of water all year. Although it has been noted that the *Lodhas* don't mind residing in mixed-caste and mixed-tribal villages, they frequently choose to establish a home amongst their own in seclusion from other groups and tribes. As a result, it is common to find the *Lodhas*' hamlets in remote, densely forested places. The *Lodhas* villages lack a clear ground layout. The secular headman of each *Lodhas* village is known as "*Pradhan*," while his subordinate is known as "*Chhatia*" or "Γ." Pradhan oversees the traditional village council and decides any disputes that are brought before it.

The *Lodhas* speak a hybrid of Bengali and corrupt Oriya 'Mundari. They have robust bodies, and their torso is muscular. The skin tone is light to medium brown, and the height is average. The *Lodhas* are physically strong and healthy, yet they are not very laborious or meticulous. All *Lodhas* visit the forest at various times of the year to gather a variety of foods and small forest products. Their primary means of subsistence in the recent past was reportedly established agriculture. However, the *Lodhas* have been forced to engage in a variety of antisocial behaviours as a result of the destruction of the forest and the expropriation of land. The majority of households currently focus mostly on farming and forest collecting, with occasional wage work as a complement. Another way to augment their little income is through the collection of tassar cocoons and the sale of firewood. Male and female counterparts work just as hard as each other. Males typically care for their children, gather fuel wood, tend cattle, harvest paddy, occasionally prepare meals, gather tassar cocoons, plough fields, and work as labourers on roads. Selling fuel wood is a significant economic activity. For capturing and hunting game animals, they have a variety of traps and snares in addition to bows and arrows. They enjoy fishing and use hand nets and basket traps to catch fish. Tortoise catching is another significant and engaging activity. The *Lodhas* once engaged in shifting farming, but they have now switched to established agriculture. Their economic activities are closely related to their agricultural cycle, forest economy, and several other sources, notably the production of plates, cash-earning leaves, and Sabai grass ropes (https://tribal.nic.in/repository/ViewDoc.aspx? RepositoryNo=TRI28-08 2017115420 and file=Docs/TRI28-08-2017115420.pdf).

2.2. *Apatani*

The *Apatani* tribe is intriguing. They are not nomadic like the other tribes. They continuously cultivate wetlands in mountainous terrain by sawing the hills, in contrast to other tribes that clear forests to cultivate the dry ground. In the enigmatic Indian state of Arunachal Pradesh, Ziro, an ancient *Apatani* Tribe settlement, may be discovered (https://www.atlasobscura.com/places/Apatani-tribal-village-ziro).

The elder women of the tribe, who have enormous nose rings and face tattoos, are arguably the most alluring. The custom, which is now extinct, originated in premodern times when *Apatani* women were regarded as the most attractive in the entire state of Arunachal Pradesh. The *Apatani* people dress elegantly and colourfully in basic traditional attire. The practice of having tattoos and inserting big nose rings, which was formerly common among women, has gradually declined in recent years. According to legend, the women started doing this to seem unappealing to men from nearby tribes. Among the Arunachal tribes, the *Apatani* women have long been regarded as the most attractive. The *Apatani* males tattooed their women's faces and made them wear big nose rings to make them less alluring since other tribes' men would frequently kidnap the *Apatani* ladies. The ladies have five vertical stripes under the lower lip on the chin and broad blue lines from the forehead to the tip of the nose tattooed on themselves. On top of their heads, the ladies fold their curls into balls (Dilling). They occasionally horizontally insert a metal skewer. Traditionally, males bind their hair in a bun just above the forehead using a twelve-inch-long horizontal metal rod. They dress in delicate cane belt strips that have been painted crimson and are twisted into the shape of horse collars with extended ends. The horse collar's loop is looped around the waist, and those cane strips are loosely linked together. The males have tattoos on their chin in the shape of a "T" behind the lower lip (http://www.easternindiatours.com/index.php?option=com_content and view= article and id=85:tribes-of-Arunachal-13nights-14days and catid=41:arunachal-pradeshtours and Itemid=62).

In the Ziro Valley in the Lower Subansiri district of Arunachal Pradesh, in the Eastern Himalayan area of India, there are seven villages where the *Apatanis* built their homes. One of the 26 significant tribes of Arunachal Pradesh is the *Apatani*. According to the 2011 Census, this established community has a total population of 29,546 people (14,583 women and 14,963 men). The whole *Apatani* population resided in the seven village clusters that make up the one valley that is now known as Ziro, the administrative centre of Lower Subansiri, until the 1950s, when the Indian government established its rule over their valley. Currently, many *Apatanis* reside outside of their native villages and in the twin cities of Ziro and Hapoli, both of which are situated in the valley (Srivastava *et al.*, 2010). Many *Apatanis* also work or attend school in other regions of the nation. Rice-cum-fish culture accounts for 48.38 per cent of the *Apatanis'* integrated productive system, which is followed by clan woods (32.64 per cent), bamboo forests (16.41 per cent), and home gardens (2.75 per cent) (Rai, 2005; Dollo *et al.*, 2009). The Mura (slave) and Mite (patrician) classes make up the bulk of the *Apatani* society. The Muras are more like mongoloid creatures, whereas the Mites have higher statures, lighter complexion, prominent noses, and deep-set eyes. The vivid green tattoo lines on an *Apatani* person's thin skin make them immediately identifiable (Bouchery, 2010). Defacing *Apatani* females with tattoos is done to deter *Nyishi* tribal members from being fascinated with them and engaging in criminal attacks on attractive *Apatani* girls in the past. Despite the patriarchal nature of the *Apatani* community, women are essential to agricultural practices and the preservation of the plateau's native biodiversity. The *Apatanis* still exclusively rely on forested plant resources for their livelihood and to sustain their farming methods in isolated areas of Arunachal Pradesh (Dollo *et al.*, 2009). The *Apatani's* indigenous traditional lifestyles depend heavily on fermented foods, which are essential and inseparable components of their diet (Tiwari and Mahanta, 2007). The *Apatani* have maintained a self-managed system of folk medicine for many generations that was mostly focused on herbal remedies.

2.3. *Kanikkar*

One of the significant tribes found in several parts of Kerala and Tamil Nadu is the *Kanikkar*. In the Indian states of Tamil Nadu, Kanyakumari, and Tirunelveli, as well as Thiruvananthapuram and Kollam in Kerala, they live in or close to the woods. *Kanis* is another name for the *Kanikkars*. The *Kanis* have a long history of hunting, collecting, and relocating agriculture, just like the other indigenous hunting and gathering tribes. The Travancore government hired the *Kanikkars* in the distant past to gather honey, wax, ginger, cardamom, dammar, and elephant tusks (Thurston, 1909). Every *Kani* settlement has a *Mootukani*, who is in charge of managing all social issues and administering justice. He will deal with the issues affecting his people, and he will preside over the sessions where problems will be solved. To tell the *Mootukani* of his directives, a *Vilikani*, or informant, gathers people in a location called Chattumadam. They speak a language that is primarily Malayalam based with a few words from Tamil (Palanichamy, 2010; Arun *et al.*, 2007). The medical professionals of the *Kani* tribe are known as Philathies (Prasad *et al.*, 1996).

The following publications are chosen for the current comparative study.

1. Chaudhury, S., Singh, H., and Rahaman, C. H. (2018). Ethnomedicinal uses of plants by the Lodhas tribal group of West Bengal, India. *Journal of Traditional and Folk Practices, 6*(1), 67-97.
2. Kala, C. P. (2005). Ethnomedicinal botany of the Apatani in the Eastern Himalayan region of India. *Journal of ethnobiology and Ethnomedicine, 1*, 1-8.
3. Rani, S. L., Devi, V. K., Soris, P. T., Maruthupandian, A., and Mohan, V. R. (2011). Ethnomedicinal plants used by Kanikkars of agasthiarmalai biosphere reserve, Western Ghats. *Journal of Ecobiotechnology, 3*(7).

3. Result and Discussion

This communication presents a comparative analysis of the indigenous medicinal plants utilised by the *Lodha, Apatani*, and *Kanikkar* tribes to treat various ailments and diseases. Different ailments endured by the *Lodha, Apatani*, and *Kanikkar* in their settlements are listed under 21 Ailment(s)/Health Categories with their biomedical terms (Table 8.1).

Table 8.1: Ailment Categories with their Biomedical Terms

Ailment(s)/ Health Categories	*Biomedical Terms*	*Total Number of Ailments*
Aches	Head Ache	01
Breastfeeding Problems	Agalactia	01
Dental and Oral Problems	Dentin Hypersensitivity, Odontalgia, and Periodontitis/Pyorrhea	03

Ailment(s)/ Health Categories	*Biomedical Terms*	*Total Number of Ailments*
Dermatological Problems	Acne, Boils, Dandruff, Dermatophytosis, Eczema (Atopic Dermatitis), Heel Fissures, Prickly Heat, Pruritus, Rashes, Scabies, Scalp Seborrheic Dermatitis, Skin Infections, and Vitiligo	13
Ear Problems	Lightheadedness/Vertigo and Otalgia	02
Eating Disorders	Anorexia	01
Follicle Disorders	Alopecia	01
Gastro-intestinal Problems	Colic, Constipation, Dysentery, Functional Dyspepsia, Gastritis, Gastroenteritis, Irritable Bowel Syndrome, and Peptic Ulcer	08
Genitourinary Problems	Dhat Syndrome/Spermatorrhea, Nephrolithiasis and Polydipsia	03
Gynecological Problems	Dysmenorrhea, Leucorrhea, Menorrhagia, and Menstrual Irregularities	04
Health Care Interventions	Abortion (Induced Miscarriage)	01
Hematological Problems	Blood Cleansing/Detoxification	03
Infections and Parasitic Diseases	Malaria, Neonatal Sepsis, Parotitis, Pneumonia, and Tonsilitis	05
Liver Problems	Hepatitis, Hepatomegaly, Hepatosplenomegaly, and Jaundice	04
Metabolic Syndrome/ Syndrome X	Diabetes and Obesity	02
Musculoskeletal System and Connective Tissue Problems	Arthritis, Fracture, Myalgia, Osteoarthritis, and Rheumatism	05
Neoplasms	Cancer	01
Ophthalmic Disorders and Diseases	Ocular Pruritis	01
Reproductive Health Problems	Impotence, Infertility, and Sexual Debility (Zo' f-e-Bah)	03
Respiratory Problems	Asthma, Cold, and Cough	06
Unspecified	Anticonception/Contraception, Burns, Disinfections, Dog Bite, Edema, General Illness, Hyperthermia, Insect Bites, Laceration, Lice Infestation/Pediculosis, Pyrexia, Scorpion Sting, and Snake Bite	13
	Total	**81**

3.1. Comparative Ethono-medico-botany of the *Lodha* in West Bengal of India and the *Apatani* in the Eastern Himalayan Region of India

A comparison between the *Lodha* and *Apatani* reveals that both of them use 19 plant species to treat 49 different human ailments. It is observed that out of 19 plant species, 9 plant species, namely *Acorus calamus* (Musculoskeletal System and Connective Tissue Problems, and Unspecified), *Andrographis paniculata* (Gastro-intestinal Problems), *Anisomeles indica* (Musculoskeletal System and Connective Tissue Problems), *Calotropis gigantea* (Unspecified), *Eclipta prostrata* (Unspecified), *Murraya koenigii* (Gastro-intestinal Problems), *Musa × paradisiaca* (Gastro-intestinal Problems), *Oxalis corniculata* (Eating Disorders), and *Terminalia chebula* (Respiratory Problems) are used by both of them to treat ailments in the same ailment categories. It is observed that out of 19 plant species, 14 plant species, namely *Acorus calamus* (Aches, Gastro-intestinal Problems, Musculoskeletal System and Connective Tissue Problems, and Unspecified), *Alstonia scholaris* (Musculoskeletal System and Connective Tissue Problems, and Unspecified), *Andrographis paniculata* (Dermatological Problems, Gastro-intestinal Problems, Infections and Parasitic Diseases, and Respiratory Problems), *Anisomeles indica* (Musculoskeletal System and Connective Tissue Problems and Unspecified), *Calotropis gigantea* (Gynecological Problems, Infections and Parasitic Diseases, and Unspecified), *Curcuma aromatica* (Gastro-intestinal Problems, Musculoskeletal System and Connective Tissue Problems, and Unspecified), *Curcuma caesia* (Dermatological Problems, Gastro-intestinal Problems, Genitourinary System, Metabolic Syndrome/Syndrome X, Musculoskeletal System and Connective Tissue Problems, Neoplasms, and Unspecified), *Dioscorea bulbifera* (Metabolic Syndrome/ Syndrome X and Musculoskeletal System and Connective Tissue Problems), *Eclipta prostrata* (Dermatological Problems, Gastro-intestinal Problems, and Unspecified), *Hibiscus rosa-sinensis* (Dermatological Problems and Gynecological Problems), *Musa × paradisiaca* (Gastro-intestinal Problems and Gynecological Problems), *Oroxylum indicum* (Infections and Parasitic Diseases and Liver Problems), *Oxalis corniculata* (Eating Disorders, Genitourinary System, and Liver Problems), and *Terminalia chebula* (Dermatological Problems, Musculoskeletal System and Connective Tissue Problems, Respiratory Problems, and Unspecified) are used by the *Lodha* to treat ailments in more than one ailment categories. It is observed that out of 19 plant species, 6 plant species, namely *Acorus calamus* (Dermatological Problems, Musculoskeletal System and Connective Tissue Problems, and Unspecified), *Alstonia scholaris* (Aches, Gastro-intestinal Problems, and Gynecological Problems), *Centella asiatica* (Gastro-intestinal Problems and Hematological Problems), *Chromolaena odorata* (Aches and Unspecified), *Oroxylum indicum* (Aches and Gastro-intestinal Problems), and *Oxalis corniculata* (Aches and Eating Disorders) are used by the *Apatani* to treat ailments in more than one ailment category (Table 8.2).

3.2. Comparative Ethono-medico-botany of the *Lodha* in West Bengal of India and the *Kanikkar* in Agasthiarmalai Biosphere Reserve, Western Ghats, Tamil Nadu

A comparison between the *Lodha* and *Kanikkar* reveals that both of them use 33 plant species to treat 63 different human ailments. It is observed that out of 33 plant species, 10 plant species, namely *Achyranthes aspera* (Unspecified), *Azadirachta indica* (Musculoskeletal System and Connective Tissue Problems), *Calotropis gigantea* (Unspecified), *Cissus quadrangularis* (Musculoskeletal System and Connective Tissue Problems), *Datura metel* (Musculoskeletal System and Connective Tissue Problems), *Eclipta prostrata* (Unspecified), and *Lawsonia inermis* (Dermatological Problems) are used by both of them to treat ailments in the same ailment categories. It is observed that out of 33 plant species, 19 plant species, namely *Abrus precatorius* (Dermatological Problems, Ear Problems, Gynecological Problems and Infections and Parasitic Diseases), *Abutilon indicum* (Aches and Gastro-intestinal Problems), *Achyranthes aspera* (Gastro-intestinal Problems, Genitourinary System, Health Care Intervention, and Unspecified), *Acorus calamus* (Aches, Gastro-intestinal Problems, Musculoskeletal System and Connective Tissue Problems, and Unspecified), *Alstonia scholaris* (Musculoskeletal System and Connective Tissue Problems and Unspecified), *Amaranthus viridis* (Gastro-intestinal Problems and Health Care Intervention), *Andrographis paniculata* (Dermatological Problems, Gastro-intestinal Problems, Infections and Parasitic Diseases, and Respiratory Problems), *Azadirachta indica* (Dermatological Problems, Musculoskeletal System and Connective Tissue Problems, and Unspecified), *Boerhavia diffusa* (Liver Problems and Musculoskeletal System and Connective Tissue Problems), *Calotropis gigantea* (Gynecological Problems, Infections and Parasitic Diseases, and Unspecified), *Carica papaya* (Gastro-intestinal Problems and Unspecified), *Coccinia grandis* (Hematological Problems, Gastro-intestinal Problems, and Unspecified), *Datura metel* (Dermatological Problems, Ear Problems, Musculoskeletal System and Connective Tissue Problems, and Unspecified), *Dioscorea bulbifera* (Metabolic Syndrome/Syndrome X and Musculoskeletal System and Connective Tissue Problems), *Eclipta prostrata* (Dermatological Problems, Gastro-intestinal Problems, and Unspecified), *Ficus benghalensis* (Gynecological Problems, Genitourinary System, and Unspecified), *Jatropha gossypiifolia* (Dental and Oral Problems, Dermatological Problems, and Health Care Intervention), *Pterocarpus marsupium* (Gynecological Problems, Metabolic Syndrome/Syndrome X, and Reproductive Health Problems), and *Tamarindus indica* (Metabolic Syndrome/Syndrome X and Reproductive and Health Problems) are used by the *Lodha* to treat ailments in more than one ailment categories. It is observed that out of 33 plant species, 3 plant species, namely *Abrus precatorius* (Dental and Oral Problems, Musculoskeletal System and Connective Tissue Problems, Ophthalmic Diseases, and Unspecified), *Alstonia scholaris* (Breastfeeding Problems and Respiratory Problems), and *Calotropis gigantea* (Musculoskeletal System and Connective Tissue Problems and Unspecified) are used by the *Kanikkar* to treat ailments in more than one ailment category (Table 8.3).

Table 8.2: Comparative Ethono-Medico-Botany of the *Lodha* in West Bengal of India and the *Apatani* in the Eastern Himalayan Region of India

Sl. No.	*Botanical Name*	*Part(s) Used*		*Ailment(s) Treated*		*Ailment(s) Categories*	
		Lodha	*Apatani*	*Lodha*	*Apatani*	*Lodha*	*Apatani*
1.	***Acorus calamus* L.** Synonym: *Acorus angustatus* Raf. Family: Acoraceae Habit: Herb	Leaf and Rhizome/ Under-ground Stem	Root	Head Aches, Gastroenteritis, Rheumatism, and Pyrexia	Acne, Boils, Dandruff, Pruritus, Prickly Heat, Fracture, and Laceration	Aches, Gastro-intestinal Problems, Musculoskeletal System and Connective Tissue Problems, and Unspecified	Dermatological Problems, Musculoskeletal System and Connective Tissue Problems, and Unspecified
2.	***Alstonia scholaris* (L.) R. Br.** Synonym: *Echites scholaris* L. Family: Apocynaceae Habit: Tree	Stem Bark, Stem Exudate (Latex), and Whole plant	Leaf and Stem Bark	Rheumatism and Snake Bite	Head Aches, Colic, Ulcer and Menstrual Irregularities	Musculoskeletal System and Connective Tissue Problems, and Unspecified	Aches, Gastro-intestinal Problems, and Gynecological Problems
3.	***Andrographis paniculata* (Burm. f.) Nees** Synonym: *Justicia paniculata* Burm. f. Family: Acanthaceae Habit: Herb	Leaf and Whole Plant	Leaf	Skin Infections, Functional Dyspepsia, Neonatal Sepsis, Cold and Cough	Dysentery	Dermatological Problems, Gastro-intestinal Problems, Infections and Parasitic Diseases, and Respiratory Problems	Gastro-intestinal Problems

Sl. No.	*Botanical Name*	*Part(s) Used*		*Ailment(s) Treated*		*Ailment(s) Categories*	
		Lodha	*Apatani*	*Lodha*	*Apatani*	*Lodha*	*Apatani*
4.	***Anisomeles indica* (L.) Kuntze** Synonym: *Ajuga disticha* (L.) Roxb. Family: Lamiaceae Habit: Herb	Leaf, Root, and Whole Plant	Shoot Apical Meristem	Myalgia, Pyrexia, Scorpion Sting, and Snake Bite	Myalgia	Musculoskeletal System and Connective Tissue Problems and Unspecified	Musculoskeletal System and Connective Tissue Problems
5.	***Calotropis gigantea* (L.) Dryand**. Synonym: *Calotropis gigantea* (L.) R. Br. ex Schult. Family: Apocynaceae Habit: Shrub	Leaf, Root, and Stem Exudate (Latex)	Root	Leucorrhea, Parotitis, Laceration, and Snake Bite	Dog Bite	Gynecological Problems, Infections and Parasitic Diseases, and Unspecified	Unspecified
6.	***Centella asiatica* (L.) Urb.** Synonym: *Hydrocotyle brasiliensis* Scheidw. ex Otto and F. Dietr. Family: Apiaceae Habit: Creeper	Leaf	Shoot Apical Meristem	Pneumonia	Constipation, Gastritis, and Blood Cleansing/ Detoxification	Infections and Parasitic Diseases	Gastro-intestinal Problems and Hematological Problems

Sl. No.	*Botanical Name*	*Part(s) Used*		*Ailment(s) Treated*		*Ailment(s) Categories*	
		Lodha	*Apatani*	*Lodha*	*Apatani*	*Lodha*	*Apatani*
7.	***Chromolaena odorata* (L.) R. M. King and H. Rob.** Synonym: *Chrysocoma maculata* Vell. Family: Asteraceae Habit: Shrub	Leaf	Leaf	Acne, Boils, Dandruff, Pruritus, and Prickly Heat	Head Aches, Laceration, and Pyrexia	Dermatological Problems	Aches and Unspecified
8.	***Curcuma aromatica* Salisb.** Synonym: *Curcuma wenyujin* Y. H. Chen and C. Ling Family: Zingiberaceae Habit: Herb	Rhizome/ Under-ground Stem	Whole Plant	Gastroenteritis, Rheumatism, and Pyrexia	Blood Cleansing/ Detoxification	Gastro-intestinal Problems, Musculoskeletal System and Connective Tissue Problems, and Unspecified	Hematological Problems
9.	***Curcuma caesia* Roxb.** Synonym: *Curcuma kuchoor* Royle Family: Zingiberaceae Habit: Herb	Rhizome/ Under-ground Stem	Rhizome/ Under-ground Stem	Eczema, Functional Dyspepsia, Nephro-lithiasis, Diabetes, Rheumatism, Cancer, and Burns	Asthma and Cough	Dermatological Problems, Gastro-intestinal Problems, Genitourinary System, Metabolic Syndrome/ Syndrome X, Musculoskeletal System and Connective Tissue Problems, Neoplasms, and Unspecified	Respiratory Problems

Sl. No.	Botanical Name	Part(s) Used		Ailment(s) Treated		Ailment(s) Categories	
		Lodha	*Apatani*	*Lodha*	*Apatani*	*Lodha*	*Apatani*
10.	***Cuscuta reflexa* Roxb.** Synonym: *Cuscuta elatior* Choisy Family: Convolvulaceae Habit: Twiner	Whole Plant	Whole Plant	Head Aches	Constipation	Aches	Gastro-intestinal Problems
11.	***Dioscorea alata* L.** Synonym: *Dioscorea atropurpurea* Roxb. Family: Dioscoreaceae Habit: Climber	Root Tuber	Root Tuber	Diabetes	Functional Dyspepsia	Metabolic Syndrome/ Syndrome X	Gastro-intestinal Problems
12.	***Dioscorea bulbifera* L.** Synonym: *Dioscorea anthropophagorm* A. Chev. Family: Dioscoreaceae Habit: Climber	Root Tuber	Root Tuber	Diabetes and Rheumatism	Functional Dyspepsia	Metabolic Syndrome/ Syndrome X and Musculoskeletal System and Connective Tissue Problems	Gastro-intestinal Problems
13.	***Eclipta prostrata* (L.) L.** Synonym: *Acmella lanceolata* Link ex Spreng. Family: Asteraceae Habit: Herb	Fruit, Leaf, and Root	Shoot Apical Meristem	Rashes, Gastritis, Gastroenteritis, Lice Infestation/ Pediculosis, and Scorpion Sting	Laceration	Dermatological Problems, Gastro-intestinal Problems, and Unspecified	Unspecified

Sl. No.	*Botanical Name*	*Part(s) Used*		*Ailment(s) Treated*		*Ailment(s) Categories*	
		Lodha	*Apatani*	*Lodha*	*Apatani*	*Lodha*	*Apatani*
14.	***Hibiscus rosa-sinensis* L.** Synonym: *Hibiscus rosa sinensis* var. *rosa-sinensis* Family: Malvaceae Habit: Shrub	Flower and Leaf	Flower	Acne, Pruritus, Dysmenorrhea, and Menorrhagia	Impotence and Infertility	Dermatological Problems and Gynecological Problems	Reproductive Health Problems
15.	***Murraya koenigii* (L.) Spreng.** Synonym: *Bergera koenigii* L. Family: Rutaceae Habit: Tree	Leaf	Leaf	Functional Dyspepsia	Functional Dyspepsia	Gastro-intestinal Problems	Gastro-intestinal Problems
16.	***Musa × paradisiaca* L.** Synonym: *Karkandela × malabarica* Raf. Family: Musaceae Habit: Herb	Fruit, Ramus/ Stem Branch, and Root	Fruit	Gastroenteritis, Dhat Syndrome/ Sperma-torrhea, Leucorrhea, and Menorrhagia	Functional Dyspepsia	Gastro-intestinal Problems and Gynecological Problems	Gastro-intestinal Problems
17.	***Oroxylum indicum* (L.) Kurz** Synonym: *Arthrophyllum ceylanicum* Miq. Family: Bignoniaceae Habit: Tree	Fruit	Seed	Malaria and Jaundice	Head Aches and Constipation	Infections and Parasitic Diseases and Liver Problems	Aches and Gastro-intestinal Problems

Sl. No.	*Botanical Name*	*Part(s) Used*		*Ailment(s) Treated*		*Ailment(s) Categories*	
		Lodha	*Apatani*	*Lodha*	*Apatani*	*Lodha*	*Apatani*
18.	***Oxalis corniculata* L.** Synonym: *Oxalis albicans* Kunth Family: Oxalidaceae Habit: Creeper	Leaf	Shoot Apical Meristem	Anorexia, Polydipsia, Hepatitis, and Jaundice	Anorexia and Head Aches	Eating Disorders, Genitourinary System, and Liver Problems	Aches and Eating Disorders
19.	***Terminalia chebula* Retz.** Synonym: *Myrobalanus gangetica* (Roxb.) Kostel. Family: Combretaceae Habit: Tree	Fruit	Fruit	Acne, Pruritus, Vitiligo, Rheumatism, Cold, Cough, and Burns	Cough	Dermatological Problems, Musculoskeletal System and Connective Tissue Problems, Respiratory Problems, and Unspecified	Respiratory Problems

Table 8.3: Comparative Ethono-Medico-Botany of the *Lodha* in West Bengal of India and the *Kanikkar* in Agasthiarmalai Biosphere Reserve, Western Ghats, Tamil Nadu

Sl. No.	Botanical Name	Part(s) Used		Ailment(s) Treated		Ailment(s) Categories	
		Lodha	Kanikkar	Lodha	Kanikkar	Lodha	Kanikkar
1.	***Abrus precatorius* L.** Synonym: *Abrus abrus* (L.) Wright Family: Fabaceae Habit: Shrub	Fruit, Root, and Stem Bark	Leaf, Root, and Seed	Acne, Boils, Dandruff, Pruritus, Prickly Heat, Otalgia, Leucorrhea, and Tonsilitis	Dentin Hyper-sensitivity, Arthritis, Ocular Pruritis, and Insect Bites	Dermatological Problems, Ear Problems, Gynecological Problems and Infections and Parasitic Diseases	Dental and Oral Problems, Musculoskeletal System and Connective Tissue Problems, Ophthalmic Diseases, and Unspecified
2.	***Abutilon indicum* (L.) Sweet** Synonym: *Sida albida* Willd. Family: Malvaceae Habit: Shrub	Leaf and Root	Leaf	Head Aches and Gastroenteritis	Otalgia	Aches and Gastro-intestinal Problems	Ear Problems
3.	***Achyranthes aspera* L.** Synonym: *Achyranthes sicula* Roth Family: Amaranthaceae Habit: Herb	Leaf and Root	Leaf	Gastroenteritis, Abortion (Induced Miscarriage), Polydipsia, Anti-conception/ Contraception, Dog Bite and Scorpion Sting	Dog Bite	Gastro-intestinal Problems, Genitourinary System, Health Care Intervention, and Unspecified	Unspecified

Sl. No.	Botanical Name	Part(s) Used		Ailment(s) Treated		Ailment(s) Categories	
		Lodha	*Kanikkar*	*Lodha*	*Kanikkar*	*Lodha*	*Kanikkar*
4.	***Acorus calamus*** **L.** Synonym: *Acorus angustatus Raf.* Family: Acoraceae Habit: Herb	Leaf and Rhizome/ Under-ground Stem	Rhizome/ Under-ground Stem	Head Aches, Gastroenterit is, Rheumatism and Pyrexia	Light-headedness/ Vertigo	Aches, Gastro-intestinal Problems, Musculoskeletal System and Connective Tissue Problems, and Unspecified	Ear Problems
5.	***Alstonia scholaris*** **(L.) R. Br.** Synonym: *Echites scholaris* L. Family: Apocynaceae Habit: Tree	Stem Bark, Stem Exudate (Latex), andWhole Plant	Leaf, Stem Bark, and Stem Exudate (Latex)	Rheumatism and Snake Bite	Agalactia andAsthma	Musculoskeletal System and Connective Tissue Problems and Unspecified	Breastfeeding Problems and Respiratory Problems
6.	***Amaranthus viridis*** **L.** Synonym: *Albersia caudata* (Jacq.) Boiss. Family: Amaranthaceae Habit: Herb	Leaf, Root, and Whole Plant	Whole Plant	Constipation and Abortion (Induced Miscarriage)	Scabies	Gastro-intestinal Problems and Health Care Intervention	Dermatological Problems
7.	***Andrographis paniculata*** **(Burm. f.) Nees** Synonym: *Justicia paniculata* Burm. f. Family: Acanthaceae Habit: Herb	Leaf andWhole Plant	Leaf and Root	Skin Infections, Functional Dyspepsia, Neonatal Sepsis, Cold and Cough	Insect Bites	Dermatological Problems, Gastro-intestinal Problems, Infections and Parasitic Diseases, and Respiratory Problems	Unspecified

Sl. No.	Botanical Name	Part(s) Used		Ailment(s) Treated		Ailment(s) Categories	
		Lodha	Kanikkar	Lodha	Kanikkar	Lodha	Kanikkar
8.	***Azadirachta indica* A. Juss.** Synonym: *Antelaea canescens* Cels ex Heynh. Family: Meliaceae Habit: Tree	Leaf, FruitRoot, Seed, and Whole Plant	Seed and Stem Bark	Acne and Pruritus, Scalp Seborrheic Dermatitis, Rheumatism, Lice Infestation/ Pediculosis, and Pyrexia	Rheumatism	Dermatological Problems, Musculoskeletal System and Connective Tissue Problems, and Unspecified	Musculoskeletal System and Connective Tissue Problems
9.	***Bacopa monnieri* (L.) Wettst.** Synonym: *Bramia monnieri* (L.) Pennell Family: Plantaginaceae Habit: Creeper	Whole Plant	Whole Plant	General Illness	Osteoarthritis	Unspecified	Musculoskeletal System and Connective Tissue Problems
10.	***Boerhavia diffusa* L.** Synonym: *Boerhavia diffusa* var. *mutabilis* R. Br. Family: Nyctaginaceae Habit: Creeper	Fruit, Leaf, Ramus/ Stem Branch, and Whole Plant	Leaf	Hepatomegaly, Hepato-splenomegaly, and Myalgia	Dermato-phytosis and Scabies	Liver Problems and Musculoskeletal System and Connective Tissue Problems	Dermatological Problems
11.	***Calotropis gigantea* (L.) Dryand.** Synonym: *Calotropis gigantea* (L.) R. Br. ex Schult. Family: Apocynaceae Habit: Shrub	Leaf, Root, and Stem Exudate (Latex)	Leaf	Leucorrhea, Parotitis, Laceration, and Snake Bite	Rheumatism and Edema	Gynecological Problems, Infections and Parasitic Diseases, and Unspecified	Musculoskeletal System and Connective Tissue Problems and Unspecified

Sl. No.	*Botanical Name*	*Part(s) Used*		*Ailment(s) Treated*		*Ailment(s) Categories*	
		Lodha	*Kanikkar*	*Lodha*	*Kanikkar*	*Lodha*	*Kanikkar*
12.	***Carica papaya* L.** Synonym: *Carica citriformis* Jacq. Family: Caricaceae Habit: Herb	Fruit	Fruit	Functional Dyspepsia, Gastritis and Pyrexia	Abortion (Induced Miscarriage)	Gastro-intestinal Problems and Unspecified	Health Care Intervention
13.	***Cissus quadrangularis* L.** Synonym: *Vitis succulenta* Galpin Family: Vitaceae Habit: Climber	Whole Plant	Tender Stem	Fracture	Rheumatism	Musculoskeletal System and Connective Tissue Problems	Musculoskeletal System and Connective Tissue Problems
14.	***Coccinia grandis* (L.) Voigt** Synonym: *Bryonia acerifolia* D. Dietr. Family: Cucurbitaceae Habit: Climber	Fruit, Leaf, and Root	Fruit	Blood Cleansing/ Detoxification, Functional Dyspepsia, Gastroenteritis, and Snake Bite	Odontalgia	Hematological Problems, Gastro-intestinal Problems, and Unspecified	Dental and Oral Problems
15.	***Commelina benghalensis* L.** Synonym: *Commelina canescens* Vahl Family: Commelinaceae Habit: Herb	Leaf	Leaf	Constipation	Scabies	Gastro-intestinal Problems	Dermatological Problems

Sl. No.	*Botanical Name*	*Part(s) Used*		*Ailment(s) Treated*		*Ailment(s) Categories*	
		Lodha	*Kanikkar*	*Lodha*	*Kanikkar*	*Lodha*	*Kanikkar*
16.	***Datura metel*** **L.** Synonym: *Brugmansia waymanii* Paxton Family: Solanaceae Habit: Herb	Fruit	Leaf	Eczema (Atopic Dermatitis), Otalgia, Rheumatism, Anti-conception/ Contraception, Burns, and Snake Bite	Rheumatism	Dermatological Problems, Ear Problems, Musculoskeletal System and Connective Tissue Problems, and Unspecified	Musculoskeletal System and Connective Tissue Problems
17.	***Dioscorea bulbifera*** **L.** Synonym: *Dioscorea anthropophagorm* A. Chev. Family: Dioscoreaceae Habit: Climber	Root Tuber	Root Tuber	Diabetes and Rheumatism	Dysentery	Metabolic Syndrome/ Syndrome X and Musculoskeletal System and Connective Tissue Problems	Gastro-intestinal Problems
18.	***Dioscorea pentaphylla*** **L.** Synonym: *Botryosicyos pentaphyllus* (L.) Hochst. Family: Dioscoreaceae Habit: Climber	Leaf and Root Tuber	Root Tuber	Acne and Pruritus	General Illness	Dermatological Problems	Unspecified

Sl. No.	Botanical Name	Part(s) Used		Ailment(s) Treated		Ailment(s) Categories	
		Lodha	*Kanikkar*	*Lodha*	*Kanikkar*	*Lodha*	*Kanikkar*
19.	***Diospyros melanoxylon* Roxb.** Synonym: *Diospyros dubia* Wall. ex A. DC. Family: Ebenaceae Habit: Tree	Fruit and Leaf	Leaf	Scorpion Sting	Arthritis and Rheumatism	Unspecified	Musculoskeletal System and Connective Tissue Problems
20.	***Eclipta prostrata* (L.) L.** Synonym: *Acmella lanceolata* Link ex Spreng. Family: Asteraceae Habit: Herb	Fruit, Leaf, and Root	Leaf and Ramus/ Stem Branch	Rashes, Gastritis, Gastroenteritis, Lice Infestation/ Pediculosis, and Scorpion Sting	Hyperthermia	Dermatological Problems, Gastro-intestinal Problems, and Unspecified	Unspecified
21.	***Ficus benghalensis* L.** Synonym: *Urostigma benghalense* (L.) Gasp. Family: Moraceae Habit: Tree	Fruit, Leaf,Prop Root, and Root	Stem Exudate (Latex)	Leucorrhea, Dhat Syndrome/ Sperma-torrhea, and Anti-conception/ Contraception	Heel Fissures	Gynecological Problems, Genitourinary System, and Unspecified	Dermatological Problems

Sl. No.	*Botanical Name*	*Part(s) Used*		*Ailment(s) Treated*		*Ailment(s) Categories*	
		Lodha	*Kanikkar*	*Lodha*	*Kanikkar*	*Lodha*	*Kanikkar*
22.	***Jatropha gossypiifolia* L.** Synonym: *Adenoropium gossypiifolium* (L.) Pohl Family: Euphorbiaceae Habit: Shrub	Fruit, Ramus/ Stem Branch, and Stem Exudate (Latex)	Stem Bark	Periodontitis/ Pyorrhea, Acne, Pruritus, and Abortion (Induced Miscarriage)	Rheumatism	Dental and Oral Problems, Dermatological Problems, and Health Care Intervention	Musculoskeletal System and Connective Tissue Problems
23.	***Justicia adhatoda* L.** Synonym: *Adhatoda zeylanica* Medik. Family: Acanthaceae Habit: Shrub	Leaf	Leaf	Cold and Cough	Rheumatism	Respiratory Problems	Musculoskeletal System and Connective Tissue Problems
24.	***Lawsonia inermis* L.** Synonym: *Lawsonia speciosa* L. Family: Lythraceae Habit: Shrub	Fruit and Leaf	Flower and Leaf	Acne and Pruritus	Dermato-phytosis and Scalp Seborrheic Dermatitis	Dermatological Problems	Dermatological Problems
25.	***Leucas aspera* (Willd.) Link** Synonym: *Leucas dimidiata* (Roth) Spreng. Family: Lamiaceae Habit: Herb	Whole Plant	Leaf	Rheumatism	Cough	Musculoskeletal System and Connective Tissue Problems	Respiratory Problems

Sl. No.	*Botanical Name*	*Part(s) Used*		*Ailment(s) Treated*		*Ailment(s) Categories*	
		Lodha	*Kanikkar*	*Lodha*	*Kanikkar*	*Lodha*	*Kanikkar*
26.	***Mallotus philippensis* (Lam.) Müll. Arg.** Synonym: *Aconceveibum trinerve* Miq. Family: Euphorbiaceae Habit: Tree	Fruit	Leaf and Tender Fruit	Disinfections	Rheumatism	Unspecified	Musculoskeletal System and Connective Tissue Problems
27.	***Murraya koenigii* (L.) Spreng.** Synonym: *Bergera koenigii* L. Family: Rutaceae Habit: Tree	Leaf	Leaf	Functional Dyspepsia	Alopecia	Gastro-intestinal Problems	Follicle Disorders
28.	***Plumbago zeylanica* L.** Synonym: *Plumbago viscosa* Blanco Family: Plumbaginaceae Habit: Herb	Leaf and Whole Plant	Root	Jaundice	Insect Bites	Liver Problems	Unspecified
29.	***Pterocarpus marsupium* Roxb.** Synonym: *Lingoum marsupium* (Roxb.) Kuntze Family: Fabaceae Habit: Tree	Leaf, Root, and Stem Bark	Ramus/ Stem Branch	Dysmenorrhea, Diabetes, and Sexual Debility (Zo'f-e-Bah)	Arthritis	Gynecological Problems, Metabolic Syndrome/ Syndrome X, and Reproductive Health Problems	Musculoskeletal System and Connective Tissue Problems

Sl. No.	Botanical Name	Part(s) Used		Ailment(s) Treated		Ailment(s) Categories	
		Lodha	Kanikkar	Lodha	Kanikkar	Lodha	Kanikkar
30.	***Santalum album* L.** Synonym: *Sirium myrtifolium* L. Family: Santalaceae Habit: Tree	Stem Bark	Secondary Xylem in the Stem (Wood)	Head Ache	Eczema (Atopic Dermatitis), Prickly Heat, and Rashes	Aches	Dermatological Problems
31.	***Tamarindus indica* L.** Synonym: *Tamarindus occidentalis* Gaertn. Family: Fabaceae Habit: Tree	Fruit and Root	Seed	Obesity and Sexual Debility (Zo'f-e-Bah)	Scorpion Sting	Metabolic Syndrome/ Syndrome X and Reproductive and Health Problems	Unspecified
32.	***Tephrosia purpurea* (L.) Pers.** Synonym: *Cracca purpurea* L. Family: Fabaceae Habit: Herb	Flower and Root	Root Bark	Head Ache	Irritable Bowel Syndrome	Aches	Gastro-intestinal Problems
33.	***Vitex negundo* L.** Synonym: *Agnus-castus incisa* (Lam.) Carrière Family: Lamiaceae Habit: Tree	Leaf	Leaf	Snake Bite	Rheumatism	Unspecified	Musculoskeletal System and Connective Tissue Problems

4. Conclusion

The *Lodha*, *Apatani*, and *Kaniakkar* (indigenous people) have a well-developed ethnomedicinal system in their settlements, where ethnomedicinal surveys have been conducted by many investigators. The Indigenous people have extensive knowledge of the flora in their surroundings. The ethnomedicinal data acquired from the comparative ethno-medico-botany of the *Lodha*, *Apatani*, and *Kaniakkar* may be useful for future pharmacological study. Pharmacologists, physicians, phytochemists, botanists, and anybody else interested in the development of alternative medicines and pharmaceuticals may find the information useful. The current study might aid in the development of novel medications based on traditional medicine for the successful treatment of many human illnesses.

5. References

1. Acharya, D. and Anshu, S. 2008. Indigenous Herbal Medicines: Tribal Formulations and Traditional Herbal Practice. Aavishkar Publishers Distributor, Jaipur, India. pp. 440.
2. Ajiwe, V. I. E., Dimonyejiaku, N., Ajiwe, A. C., Chinweuba, A. J. and Chendo, N. M. 2008. Preliminary Study on the Pharmaceutical Constituents of *Emilia sonchifolia* Leaf. *Anal. Chem. J.* **2**: 302-309.
3. Alcorn, B. J. 1984. Huastec Mayan Ethnobotany. University of Texas Press, Austin, USA.
4. Arun, V., Liju, V. B., Reena, J. V., Parthipan, B. and Renuka, C. 2007. Traditional Remedies of Kani Tribes of Kottoor Reserve Forest, Agastyavanam, Thiruvananthapuram, Kerala. *Indian J. Tradit. Know.* **6**: 589-94.
5. Balick, M. J. 1996. Transforming Ethnobotany for the New Millennium. *Ann. Missouri Bot. Gard.* **83**: 58-66.
6. Bannerman, R. H. 1982. Traditional Medicine in Modern Health Care, World Health Forum. **3**: 8-13.
7. Baskey, Dhirendra Nath. *Paschim Banger Adibasi Samaj.* Subarnarekha, 1987.
8. Bishagratna, K. K. 1907-1916. An English Translation of Sushruta Samhita. Calcutta.
9. Biswal, G. C. "Tribal Development and Education: Some Issues." *Vanyajati,* vol. 32, no. 1, 1985, pp. 14.
10. Bouchery, P. 2010. The Apatani or Tanni Lecturer in Social Anthropology, France, (http: //knol.google.com/k/the-apatanior-tanii#).
11. Bruce, J. and Meeus, C. 2002. Readers Digest, Sydney.
12. Buenz, E. J., Schnepple, D. J., Bauer, B. A., Elkin, P. L., Riddle, J. M. and Motley, T. J. 2004. Techniques: Bioprospecting Historical Herbal Texts by Hunting for New Leads in Old Tomes. *Trends Pharmacol. Sci.* **25**: 494-498.

13. Carrió, E. and Vallès, J. 2012. Ethnobotany of Mallorca (Balearic Islands): A Multidisciplinary Approach. *Coll. Antropol.* **3**: 1027-1032.

14. Cotton, C. M. 1996. Ethnobotany: Principles and Applications. Chichester, UK: John Willey and Sons Ltd. pp. 424.

15. Dalziel, J. M. 1961. The Useful Plants of West Tropical Africa, the Crown Agents, London. pp. 308.

16. Dollo, M., Samal, P. K., Sundriyal, R. C., and Kumar, K. 2009. Environmentally Sustainable Traditional Natural Resource Management and Conservation in Ziro Valley, Arunachal Himalaya, India. *Am. J. Sci.* **5**: 41-52.

17. Dutta, B. K. and Dutta, P. K. 2005. Potential of Ethnobotanical Studies in North East India: An Overview. *Indian J. Tradit. Know.* **4**: 7-14.

18. Elwin, V. 1955, A Philosophy of NEFA, Shillong,: Sachin Roy and Co.

19. Fransworth, N. R. 1988. Screening Plants for New Medicines. In: Wilson, E. O. (Ed.) Biodiversity. Washington DC: National Academy Press. pp. 83-97.

20. Fransworth, N. R. 1994a. Ethnopharmacology and Drug Development. In: Wiley Chichester (Ciba Foundation Eds.) Ethnobiology and the Search for New Drugs. Chicago, USA. pp. 42-59.

21. Fransworth, N. R. 1994b. Ethnopharmacology and Drug Development. Cipa Found Symp. **185**: 42-51.

22. Getahun, A. 1976. Some Common Medicinal and Poisonous Plants Used in Ethiopian Folk Medicine. Addis Ababa University, Ethiopia. pp. 63.

23. Gurib-Fakim, A. 2006. Medicinal Plants: Traditions of Yesterday and Drugs of Tomorrow. *Mol. Aspects Med.* **27**: 1-93.

24. Hiremath, V. T. and Taranath T. C. 2013. Indigenous Traditional Knowledge on Tree Species Survey of Jogimatti Forest, Chitradurga District, Karnataka, India. *J. Ethnobiol. Tradit. Med.* **118**: 222-227.

25. http: //inet.vidyasagar.ac.in: 8080/jspui/bitstream/123456789/5449/17/17_ Chapter per cent 206.pdf

26. http: //www.easternindiatours.com/index.php?option=com_content and view=article and id=85: tribes-of-Arunachal-13nights-14days and catid=41: arunachal-pradesh-tours and Itemid=62

27. https: //www.atlasobscura.com/places/Apatani-tribal-village-ziro

28. Hussain, K., Shahzad, A. and Hussnain, S. Z. 2008. An Ethnobotanical Survey of Important Wild Medicinal Plants of Hattar District Haripur, Pakistan. *Ethnobot. Leaflets.* **12**: 29- 35.

29. Jain, D. L., Baheti, A. M., Jain, S. R. and Khandelwal, K. R. 2010. Use of Medicinal Plants among Tribes in Satpuda Region of Dhule and Jalgaon Districts of Maharashtra - An Ethnobotanical Survey. *Indian J. Tradit. Know.* **9**: 152-157.

30. Jain, S. K. 1994. Ethnobotany and Research of Medicinal Plants in India. In: Chadwick, D. J. and Marsh, J. (Eds.) Ethnobotany and the Search for New Drugs (Ciba Foundation Symposium 185). John Wiley and Sons Ltd., Chichester.

31. Jain, S. K. 1996. Glimpses of Indian Ethnobotany, Oxford and IBH Publishing Co. New Delhi, Bombay, Calcutta.

32. Jain, S. K. and Mudgal, V. 1999. A Handbook of Ethnobotany. Bishen Singh Mahendra Pal Singh Publications, Dehra Dun, India.

33. Jansen, P. C. M. 1981. Spices, Condiments and Me. dicinal plants in Ethiopia, Their Taxonomy and Agricultural Significance. Center for Agricultural Publishing and Documentation, Wageningen, Netherlands. pp. 327.

34. Jeyaprakash, K., Ayyanar, M., Geetha, K. N. and Sekar, T. 2011. Traditional Uses of Medicinal Plants among the Tribal People in Theni District (Western Ghats), Southern India. *Asian Pac. J. Trop. Biomed.* **1**: 20-25.

35. Kala, C. P. 2005. Current Status of Medicinal Plants Used by Traditional Vaidyas in Uttaranchal State of India. *Ethnobot. Res. Appl.* **3**: 267-278.

36. Kibebew, F. 2001. The Status and Availability of Oral and Written Knowledge on Traditional Health Care in Ethiopia. In: (Medhin Zewdu and Abebe Demissieeds). Conservation and Sustainable Use of Medicinal Plants in Ethiopia. Proceeding of the National Workshop on Biodiversity Conservation and Sustainable Use of Medicinal Plants in Ethiopia. *IBCR*. Addis Ababa. pp. 107-119.

37. Kirtiker, K. R. and Basu, B. D. 1993. Indian Medicinal Plants. 2nd ed. I. Page XXIII, Allahbad, India. pp. 74-120.

38. Mahmood, A., Mahmood, A. and Tabassum, A. 2011. Ethnobotanical Survey of Plants from District Sialkot, Pakistan. *J. APP. Pharm.* **2**: 212-220.

39. Majumdar, D. N., and Madan, T. N. *An Introduction to Social Anthropology.* Asia Publishing House, 1980.

40. Martin, G. J. 1995. Ethnobotany: A Methods Manual. Chapman and Hall, London. pp. 265-270.

41. Mukul, S. A., Uddin, M. B. and Tito, M. R. 2007. Medicinal Plant Diversity and Local Healthcare among the People Living in and around a Conservation Area of Northern Bangladesh. *IJFUM.* **8**: 50-63.

42. Nithya, N. R. 2014. Globalization and the Plight of Tribals: The Case of Kerala, India. *Dawn Journal.* **3**. 727-758.

43. Palanichamy, C., Mohan, V. R. and Kalidas, C. 2010. An Ethnobotanical Survey of Plants Used by the Kanikkar Tribe of Kanyakumari District Of Western Ghats, Tamil Nadu For The Treatment Of Skin Diseases. *J. Herbal. Med. Toxicol.* **4**: 179-190.

44. Panda, Nishakar. *Policies, Programmes and Strategies for Tribal Development: A Critical Appraisal.* Kalpaz Publications, 2006.

45. Park, E. J. and Pezzutto, J. M. 2002. Botanicals in Cancer Chemoprevention. *Cancer Metastasis Rev.* **21**: 231-255.

46. Parvathamma, C. *Scheduled Caste and Tribes: A Soio-Economic Survey.* Asish Publishing House, 1984.

47. Patil, D. A. 2008. Traditional Knowledge: Views, Necessity and Prospects, In: Herbal Cures: Traditional Approach, Aavishkar Publishers, Jaipur, India.

48. Posey, D. 1999. Ethnobotany: Its Implication and Application. In: Posey, D.A. and Overol, W. L. (Eds.) Proceeding of the First International Congress of Ethnobiology. **1**: 1-7.

49. Prasad, N. P., Natrajan, C. R., Narayanan, L. M. and Ranjith Singh, A. J. A. 1996. Ethnobotany of Kannikkars of South Tamil Nadu. *Indian J. Econ. Taxon. Bot.* **12**: 292- 298.

50. Pushpangadan, P. and Atal, C. K. 1984. Ethno-medico-botanical Investigations in Kerala. Some Primitive Tribes of Western Ghats and Their Herbal Medicines. *J. Ethnopharmacol.* **11**: 59-77.

51. Quanash, N. 1998. Bicultural Diversity and Integrated Health Care in Madagascar. *Nat. Resour.* **30**: 18-22.

52. Rai, S. C. 2005. Apatani paddy-cum Fish Cultivation: An Indigenous Hill Farming System of North East India. *INDIAN J. TRADIT. KNOWL.* **4**: 65-71.

53. Rai, S. C. 2005. Apatani paddy-cum Fish Cultivation: An Indigenous Hill Farming System of North East India. *INDIAN J. TRADIT. KNOWL.* **4**: 65-71.

54. Rajiv, S. K. 1998. Ethnobotany: The Renaissance of Traditional Herbal Medicine, INA Publishers, Jaipur.

55. Rashkow, E. 2023. The Nature of Endangerment in India: Tigers, 'Tribes', Extermination and Conservation, 1818-2020 Oxford Academic Books, Oxford University Press.

56. Risley, Sir Herbert. *The People of India.* Thecker, Sprink and Company, 1908.

57. Sajem, A. L. and Gosai, K. 2006. Traditional Use of Medicinal Plants by the Jaintia Tribes in North Cachar Hills District of Assam, Northeast India. *J. Ethnobiol. Ethnomed.* **2**.

58. Savnur, H. C. 1993. A Handbook of Ayurvedic Materia Medica, VI, Dr. Jarthar and Sons, Maruthi Street, Belgaum, Karnataka, India. pp. 258-261.

59. Schultes, R. E. 1962. The Role of Ethanobotanists in the Search of New Medicinal Plants, Lloydia. **25**: 257-266.

60. Schultes, R. E. and Reis, S. 1995. Ethnobotany: Evolution of a Discipline. Portland: Dioscorides Press.

61. Shroff, J. 1997. Forest Policy and Tribal Development. In: Forest Management in Tribal Areas (Eds. Mohapatra, P. M. and Mohapatra, P.) Concept Publishing Company, Mohan Garden, New Delhi, India.

62. Silva, T. D. 1997. Industrial Utilization of Medicinal Plants in Developing Countries. In: Medicinal Plants for Forest Conservation and Health Care. (Eds. Bodeker, G., Bhat, K. K. S. Burley, J. and Vantomme, P.) FAO Non-wood Forest Products Series No. 11. FAO, Rome, Italy. pp. 34-44.

63. Singh, L.R., 1965. The Tarai Region of U.P.: A study in Human Geography. Ram Narain Lal Beni Prasad, Allahabad, India, Pages: 145.

64. Singh, M., Walia, S. and Singh, A. K. 2013. Phytochemical Characterization and Antibacterial Activity of *Lindenbergia indica* Vatke. A Common Wall Flora against Some Human Pathogens in Doon Valley, Uttarakhand. *Indian J. Nat. Prod. Resour.* **4**: 138-145.

65. Solecki, R. and Shanidar, I. V. 1975. A Neanderthal Flower Burial in Northern Iraq. *Science*. **190**: 880- 881.

66. Srivastava, R. C., Singh, R. K., Apatani Community, and Mukherjee, T. K. 2010. Indigenous Biodiversity of Apatani Plateau: Learning on Biocultural Knowledge of Apatani Tribe of Arunachal Pradesh for Sustainable Livelihoods. *INDIAN J. TRADIT. KNOWL*. **9**: 432-442.

67. Staff 2007. Ethnobotany. New Zealand. *J. Bot*. **21**.

68. Thomas, H. 1995. Indigenous Knowledge, Emanicipation and Alination. *J. Know. Tf. Utlzn*. **8**: 63-73. University of Washington.

69. Thurston F and Rangachari K., **1909**: Castes and Tribes of southern India. Vol, IV. Government press.

70. Tiwari, S. C. and Mahanta, D. 2007. Ethnological Observations on Fermented Food Products of Certain Tribes of Arunachal Pradesh. *INDIAN J. TRADIT. KNOWL*. **6**: 106 -110.

71. United Nations University Institute of Advanced Studies (UNU-IAS). 2013. In: Traditional Knowledge and Biodiversity. Payyappallimana, U. and Fadeeva, Z. (Eds). Yokohama, Japan: UNU-IAS. pp. 8-9.

72. World Health Organization (WHO) 2002-2005. Geneva: Traditional Medicine Strategy.

73. World Health Organization (WHO) 2008. Traditional Medicine: Definitions. pp. 04-20.

74. Yan, X., Murphy, B. T., Hammond, G. B., Vinson, J. A., and Nieto, C. C. 2002. Antioxidant Activities and Antitumor Screening of Extracts from Cranberry Fruit (*Vaccinium macrocarpon*). *J. Agric. Food Chem*. **50**: 5844-5849.

Chapter 9

Comparative Ethno-Medico-Botany of the *Mandi* Ethnic Community in Bangladesh with Some Other Ethnic Communities in India

N. Ahamed Kabir[1], G. Bhavani[2] and A. Doss[3]

[1]Research Scholar, PG and Research Department of Botany, V.O. Chidambaram College, Thoothukudi
[2]PG and Research Department of Botany, V. O. Chidambaram College, Thoothukudi

ABSTRACT

In the present study the traditional medicinal plants used for treating various human ailments by the Mandi community of Bangladesh are compared with the traditional medicinal plants used by the Khandas in Kandhamal District of Orissa, Paliyans of Sirumalai Hills in Dindigul District of Tamil Nadu, Pawra of Satpura Hills, Maharashtra, Bhotiya Community in Central Himalaya, India to know about the uniformity and variation in ailments treated, part(s) used, use of fresh, dry plant part(s), use of a single or a combination of plants for remedy preparation, mode of remedy preparation, dose requirement, and duration. The ethnomedicinal information of 22 plant species as having medicinal properties against about 38 human ailments under 11 categories was gathered from 5 research papers were chosen for this comparative study. Since the areas colonized by Mandi, Kandhas, Paliyans, Pawra, and Bhotiya have the good ethno-botanical potential for traditional medicinal plants, they could be suitable places for further comparative Ethno-medico-botany studies and ethno-

pharmaceutical studies. This is the first comparative Ethno-medico-botany of the Mandi settled in Bangladesh with the Kandhas, Paliyans, Pawra, and Bhotiya settled in India has not yet been followed by researchers in the field of Ethnobotany.

Keywords: *Ethno-medico-botany, Mandi community, Khandas.*

1. Introduction

1.1. Flora of India

Plants, from time immemorial, have served human beings as sources of food, shelter, clothing, and medicines. Before the advent of modern allopathic medicine and synthetic drugs, plants and to a certain extent, animals and minerals were used in various formulations for the treatment of diseases by traditional medicinal practitioners. (Cotton, 1996) From the world, 420,000 flowering plants have been reported and more than 50,000 plants have been used for medicinal purposes. (Govaerts, 2001 and Schippman *et al.,* 2002) India is rich in its flora. There are 17,500 species of higher plants, 64 gymnosperms, 1200 pteridophytes, 2850 bryophytes, 2021 lichens, 15,500 fungi, and 6500 algae are reported in India in which 5,725 Angiosperms, 10 gymnosperms, 193 pteridophytes, 678 bryophytes, 466 lichens, 3500 fungi, and 1924 algae are endemic. (Sanjappa, 2005). Traditional medical knowledge of medicinal plants and their uses by indigenous cultures is not only useful for the conservation of cultural traditions and biodiversity but also for community healthcare and drug development in the present and future. (Pei, 2001).

1.2. Traditional Medicine

Throughout history, plants have been used by human beings for medicinal purposes and even in modern times have formed the basis of many pharmaceuticals in use (Schmidt, *et al.,* 2008). Plants produce a vast array of secondary metabolites as a defense against environmental stress or other factors like pest attacks, wounds, and injuries. The complex secondary metabolites produced by plants have found various therapeutic uses in medicine from time immemorial. The early history of modern medicine contains descriptions of plant-derived phytochemicals, many of which are still in use. Some examples are the discovery of cardiotonic in foxglove, salicylic acid in willow bark, and morphine in poppies (Rishton, 2008). Various terpenoid compounds, which are synthesized in plants as secondary metabolites are proving their potential in modern scientific studies against inflammatory diseases and cancer (Salminen, 2008). A recent review has shown that approximately 25 per cent of modern medications have been plant-derived, while 75 per cent of new drugs against infectious diseases that have arrived between 1981 and 2002 originated from natural sources (Bedoya, *et al.,* 2009). Herbs and spices have been used from ancient times not only to increase palatability in foods but also for their health-beneficial effects. Scientific studies have demonstrated the anti-oxidative potential of many herbs and spices, which is important since oxidative damages have been implicated in major diseases like atherosclerosis, arthritis, and diabetes. The cholesterol-lowering effect of garlic, the use of ginger in ameliorating arthritic knee pain, and the effect of several herbal supplements on psychological and cognitive function have been reviewed by Tapsell *et al.,* (2006). Despite the progress in modern medicine, it has been reported that more than 70 per cent of the developing world's population still depends on complementary and alternative systems of medicine, otherwise known as traditional medicine (Shaikh, 2005).

World Health Organization reported that the traditional medical practitioners treat about 85 per cent of patients in India, which shows the significant valuable relation between local tribal people with medicinal plants. The medicinal uses of plants are very old. The writings indicate that the therapeutic use of the plant is being practiced since 5000-4000 B.C. and the Chinese used the first natural herbal preparations as medicines. In India, however, the earliest references of the use of plants as medicine appear in Rig-Veda which is said to be written between 3500-1600 B.C. Later the properties and therapeutic uses of medicinal plants were studied in detail and recorded empirically by the ancient physician in Ayurveda which is a basic foundation of ancient medical science in India2a. Traditional medicine studies include ethnomedicine, which involves the practices most used by people that live in rural areas and indigenous communities, and this ethnomedicine is affected by factors such as age, gender, economic activity, socio–economic level, migration, access to new healthcare systems, new herbal products and urbanization [Perez-Nicolas *et al.,* 2017, Bena Bolesselem *et al.,* 2017, Srithi *et al.,* 2009]

1.3. Ethnomedicine

Ethnomedicine may be defined broadly as the use of plants by humans as medicines, but these uses could be called, more accurately, ethnobotanical medicine (Fransworth, 1994). Ethnomedicine is one of the systems of medicine that is widely practiced among the tribal and aboriginal populations of our country for the treatment of ailments (Singh *et al.,* 2013) Ethnomedical information/ data are playing an important role in developing new scientifically validated and standardized drugs, *i.e.* both herbal and modern (Savnur,1993). Ethnobotanists can play a very useful role in conserving ethnobotanical knowledge as part of living cultural, ecological systems helping to maintain a sense of pride in local cultural knowledge and practice and reinforcing links among communities and the environment, so essential for conservation (Martin, 1995).

1.4. Ethnobotany

Ethnobotany is the study of the relationship between plants and people: From "ethno"- study of people and "botany" - the study of plants (Staff, 2007). Ethnobotany is totally in virtually a new field of research if, in this field plants investigated thoroughly and systematically, it will yield results of the great value of the archeologists, anthropologists, plant geographers, ethnobotanist, linguistics, botanists, and phytochemists. After the time of Harshberger (1896) to the present date, several authors have tried to describe subject ethnobotany and its scope, methodology, its various disciplines, sub-disciplines, and potential, *etc.* (Nitesh Kumar *et al.*2014) Ethnomedicinal knowledge is also important from a humanitarian point of view in that in the long run as this knowledge may help to identify important medicinal uses that can help in curing and healthcare around the world. Attempts should be made to share the benefits arising from such knowledge with its holders. (Chandra Prakash, 2005) Scientists have now re-focused their interests on the traditional medicinal systems of many countries in their efforts to find out newer drugs for emerging diseases and better drugs for already existent diseases. Traditional medicinal systems, merely because of their long usage, have developed extensive knowledge on the various properties of different plants and plant parts and which properties, if efficiently utilized, can lead to many novel and effective drugs (Rahmatullah *et al.,* 2011)

1.5. Tribals in India

India possesses a total of 427 tribal communities, of these more than 130 major tribal communities live in North East India, which is comprised of the 8 states Meghalaya, Mizoram, Manipur, Tripura, Sikkim, Assam, Nagaland, and Arunachal Pradesh. (Ramakrishnan, 1992; Dutta *et al.,* 2005) Every tribal group represents unique indigenous ethnobotanical systems that include the mode of taking or applying externally or internally plant parts as a cure. Without proper documentation of such knowledge, the cultural and traditional heritage of the tribe loses its importance and traditional indigenous knowledge is being lost. Tribal communities are forced to change their livelihood which leads to ethnocultural degradation (Sanjeet Kumar *et al.,* 2012). Most ethnobotanical studies have been restricted to studies of tribal and rural people to record their knowledge and use of plants and to search for new sources of herbal drugs, edible plants, and other properties of plants of value to man (Jain and Mudgal, 1999). From the study of the anthropological literature, it appears that the term tribe has been used for those groups of human beings, whose place of residence is situated in remote areas like hills, forest, sea coasts, and islands; and whose style of life is quite different from the present-day civilized men. The forest occupies a central position in tribal culture and economy (Nithya, 2014). Almost every section of the Indian population uses plants as medicine and altogether about 7500 species of plants are being used by several ethnic communities. Particularly, tribal people collect and preserve locally available wild and cultivated plant species and practice herbal medicine to treat a variety of diseases and disorders. (Kala, 2005).

1.6. Tribal Medicine

Tribal medicine or traditional medicine plays a vital role in the primary healthcare of tribal as well as rural people (Rajiv, 1998; Patil, 2008). Traditional medicine which is widespread throughout the world has been recognized by the World Health Organization (WHO) as an essential building block of primary healthcare (Bannerman, 1982). Today, these traditional medicines are increasingly popular as a cost-effective alternative to, or complementary to, orthodox medicine

(Tang and Haliwel, 2010). Nowadays, traditional medicinal practices form an integral part of complementary or alternative medicine. Although their efficacy and mechanisms of action have not been tested scientifically in most cases, these simple medicinal preparations often mediate beneficial responses due to their active chemical constituents (Park and Pezzutto, 2002). About 85 per cent of traditional medicines are plant-derived (Fransworth, 1988), and about 80 per cent of the population in developing countries still use plant-based traditional medicine for their healthcare (Silva, 1997). This reliance is much frequent in the case of tribal communities as their livelihoods are mainly dependent on forests that have also built up their socioeconomic and cultural life (Shroff, 1997; Mukul *et al.,* 2007).

1.7. Cross-Cultural Studies

Cross-cultural studies were parsimoniously studied before 1998 (Chandra Prakash, 2005), but nowadays, they are increasingly being analyzed (Pieroni *et al.*2011; Saslis *et al.,* 2011: Lardos and Heinrich, 2013; Gairola *et al.,* 2014; Aziz *et al.,* 2018). Intercultural comparison has practical applications because we can address both the consensus and variations of plant use knowledge (Heinrich *et al.*1998). Rao (2015) strongly stresses the need for shortlisting and prioritizing the leads for a specific ailment by crisscrossing of information through cross-cultural studies among different ethnic tribes within a country and then compare with other developing countries for intense bioprospecting and then product development. It is believed that the use of a particular species for the same ailment by different unrelated ethnic groups certainly indicates the efficacy and potential of these plants for drug development. Biodiversity prospecting, particularly on traditional medicinal plants can certainly result in some lead/novel molecules of great economic significance.

2. Materials and Methods

In the present study the traditional medicinal plants used for treating various human ailments by the *Mandi* community of Bangladesh are compared with the traditional medicinal plants used by the *Khandas* in Kandhamal District of Orissa, *Paliyans* of Sirumalai Hills in Dindigul District of Tamil Nadu, *Pawra* of Satpura Hills, Maharashtra, *Bhotiya* Community in Central Himalaya, India to know about the uniformity and variation in ailments treated, part(s) used, use of fresh, dry plant part(s), use of a single or a combination of plants for remedy preparation, mode of remedy preparation, dose requirement and duration.

2.1. *Mandi*

The *Mandis* or Garos are an indigenous Tibeto-Burman ethnic group from the Indian subcontinent,notably found in the Indian states of Meghalaya, Assam, Tripura, Nagaland, and neighbouring areas of Bangladesh. https://www.ethnologue.com/language/grt). The *Mandi* language has some similarities with Boro-Kachari, Dimasa, and Kok-Borok languages. *Mandis* have their weapons. One of the principal weapons is a two-edged sword called *mil·am* made of one piece of iron from hilt to point. The other types of weapons are shield, spear, bow and arrow, ax, dagger, *etc.* The staple cereal food is rice. They also eat millet, maize, tapioca, *etc. Mandis* are very liberal in their food habits. They rear cows, goats, pigs, fowls, ducks, etc., and relish their meat. They eat other wild animals like deer, bison, wild pigs, *etc.* Fish, prawns, crabs, eels, and dry fish are a part of their food. The forests provide them with vegetables and roots for their curry. Bamboo shoots are esteemed as a delicacy. They use a kind of potash in curries, which they obtain by burning dry pieces of plantain stems or young bamboo locally known as *kalchi* or *katchi*. After they are burnt, the ashes are collected and dipped in water; they are strained in conical shapes in a bamboo strainer. These days most of the townspeople use cooking soda from the market in place of ash water. The *Mandis* make their liquor by fermenting a special type of rice and the finished product is called "Minil Bichi". Besides other drinks, country liquor plays an important role in the life of the *Mandis*. (https://en.wikipedia.org/wiki/Garo people)

2.2. *Kandhas*

The ***Kandha*** or ***Dongria Kondh*** people are members of the Kondhs. They are located in the Niyamgiri Hills in the state of Odisha (formerly Orissa) in India. The people of Niyamgiri use Kui language. Kui language is not written, but it is spoken among the people of the *Kondh* community. The *Dongria Kondh* community is also known as the *Kutumba*. To make the work of the body more efficient the *Kutumba* is then divided into two groups one which functions at the level of a clan or *Kuda Kutumba* and one at the level of settlement or *Nayu Kutumba*. The management of each clan is then done according to a more root level division, in which a separate group of people is dedicated to addressing the religious and political matters of the *Dongria Kondh* hence four functional groups or *punjas* are formed. The *punjas* of the *Dongria Kondhs* are *jani, pujari, bismajhi,* and *mandal*. (https://en.wikipedia.org/wiki/Dangaria Kandha)

2.3. *Paliyans*

It is believed that the *Paliyans* are indigenous people of the Palani Hills. In the Palani Hills, they are found at an altitude of up to 2200 m. Generally, the *Paliyans* are illiterate and they speak Tamil (mother tongue of Tamil Nadu). They are traditional Nomadic hunter-gatherers, honey hunters, and foragers. Yams are their major food source. In the early part of the 20th century, the *Paliyans* dressed scantily and lived in rock crevices and caves. Most have now have transformed into traders of forest products, food cultivators, and beekeepers. (Shenbaham, 2016). The *Paliyans,* when compared to various tribal communities in Tamil Nadu constitute relatively a small group (Ignacimuthu *et al.,* 2006). The *Paliyans* are also found in the hilly regions of Madurai, Dindigul, Theni, Tirunelveli, and Virudhunagar Districts. Reportedly the descendants of *Paliyans* still live in Kukul cave in Kodaikanal set high in the Palani Hills at 2100 m. *Paliyans* used to live in small caves or a hut-like structure between two rocks. Still, now some *Paliyans* live in rocky caves while roaming in the interior forest. Some who are living in the interior forest build huts with a minimum of wood and grass (Shenbaham, 2016). Traditional botanical preparations represent the primary resource for health care in *Paliyan* society. Tribal herbal practitioners play an important role in helping keep the *Paliyans* healthy. The most common ailments treated are skin problems, burns, wounds, cuts, and diseases of the skin. Other illnesses alleviated by herbal medicines include respiratory infections, coughs, fevers, and colds, and gastrointestinal problems, abdominal pains, stomach aches, throat infections, snake bites, nervous disorders, and more (Ignacimuthu *et al.,* 2006).

2.4. *Pawra*

The *Pawra* is one of the major tribes in Maharashtra. Their original habitat is the former Udaipur state in Rajasthan. (Prasad, 1996) *Pawras* usually reside in remote areas of the forests and are indiscriminately scattered in small villages known as 'padas' typically of 2 to 10 houses (Fig. 2). The distance between each 'pada' is about 1 km, a group of which forms a village. In some places, villages are of closely situated houses surrounded by farmland. Houses of *Pawra* are typical, low costing and facing north-south, constructed by using locally available material. They are mainly dependent on forest and agricultural produce. Major occupations are a collection of plant parts like fruit, edible tubers, gums, nuts, and leafy vegetables, *etc.* from the forest. In the hilly area, agriculture is completely dependent on rainfall whereas, in plains, where the facility is available, irrigated crops are also cultivated. Since this tribe usually resides in the interior and hilly regions, they are more prone to suffer from many ailments because of poor nutrition, and starvation. The habit of smoking and chewing tobacco and consuming local liquor is very common. The frequency of suffering from skin diseases like scabies, eczema, fungal infections, *etc.* is quite high. Waterborne diseases like diarrhea and dysentery are also common. The accessibility of health centers in some areas is difficult hence, the rate of mortality due to curable diseases is high. The usual tendency is to avoid going to the primary health center and reliance on traditional practices is common. Snakebite, insect bites, and attacks on wild animals are common causes of accidental deaths. (Jagtap *et al.,* 2009)

2.5. *Bhotiya*

Bhotiya or *Bhot* are groups of ethno-linguistically related Tibetan people living in the Trans-Himalayan region that divides India from Tibet. The Bhotiya speak numerous languages including Ladakhi. *Bhotiya* have six recognizable sub-groups: the *Bhot, Bhotiya,* the *Bhutia* of Sikkim, the *Tibbati* (of Sikkim and Arunachal Pradesh), the *Bhut,* the *Gyakar Khampa* of Khimling, *Bhidang* of Uttarakhand. *Bhotiya* tribe are also natives of other countries outside India and they are in Pakistan, Afghanistan, Kazakhstan, Nepal, Bhutan, and Tibet. (https://en.wikipedia.org/wiki/Bhotiya)The *Bhotiya* have their own culture, tradition, and religious beliefs. Their major occupations have been sheep rearing and agriculture, with agriculture taking precedence over pastoralism at present. Almost all of the households are involved in agricultural activities by way of subsistence farming.

(Phondani *et al.*2010)

The following research papers are chosen for this comparative study.

i. Partha, P. and Hossain, A.E., 2007. Ethnobotanical investigation into the *Mandi* ethnic community in Bangladesh. *Bangladesh Journal of Plant Taxonomy*, **14(2)**: 129-145

ii. Behera, S.K., Panda, A., Behera, S.K. and Misra, M.K., 2006. Medicinal plants used by the *Kandhas* of Kandhamal district of Orissa. *Indian Journal of Traditional Knowledge*.**5 (4)**: 519-528

iii. Karuppusamy, S., 2007. Medicinal plants used by *Paliyan* tribes of Sirumalai hills of southern India. *Natural Product Radiance*, **6(5)**: 436-442

iv. Jagtap, S.D., Deokule, S.S., Pawar, P.K. and Harsulkar, A.M., 2009. Traditional ethnomedicinal knowledge is confined to the *Pawra* tribe of Satpura Hills, Maharashtra, India. *Ethnobotanical Leaflets*, **13**: 98-115

v. Phondani, P.C., Maikhuri, R.K., Rawat, L.S., Farooquee, N.A., Kala, C.P., Vishvakarma, S.C.R., Rao, K.S. and Saxena, K.G., 2010. Ethnobotanical uses of plants among the Bhotiya tribal communities of Niti Valley in Central Himalaya, India. *Ethnobotany Research and Applications*, **8**:233-244

3. Results and Discussion

Comparative Ethno-medico-botany of the *Mandi* in Bangladesh with the *Kandhas* in Kandhamal District of Orissa Different human ailments faced by the *Mandi, Kandhas, Paliyans, Pawras,* and the Sub-communities: *Tolchha* and *Marcha* of the *Bhotiya* in their settlements are grouped under 11 ailment categories with their biomedical terms (Table 9.1).

Table 9.1: Different Ailments Reported in the Research Paper Chosen for Comparative Study Grouped under different Ailment Categories/ Organ System Categories with their Biomedical Terms

Biomedical Terms for Ailments Ailment Categories/Organ System	*Categories*
Bilious Fever, Body Ache, Hyperpyrexia, Intermittent Fever, Malaria, Nociceptive Pain, Pyrexia, and Typhoid	Aches and Fever
Diarrhoea, Dysentry, Indigestion, Intestinal Worm Infestation and Piles,	Digestive System
Canker Sore, Gingivitis, and Oral Problems	Dental and Oral System
Diabetes	Endocrine System
Angioedema, Animal Bites, Candidiasis, Common Skin Diseases, Dermatitis, Itches, Leprosy, Poisonous Bites, Wounds and Vitiligo	Integumentary System
Jaundice and Hepatosplenomegaly/Liver Pain	Liver System

3.1. Comparative Ethno-medico-botany of the *Mandi* in Bangladesh with the *Paliyans* of the Sirumalai Hills in Dindigul District of Tamil Nadu

Comparison between the *Mandi* and the *Kandhas* (Table 9.2) reveals that both of them use 13 plant species, namely *Abutilon indicum, Achyranthes aspera, Andrographis paniculata, Asparagus racemosa, Cissus quadrangularis, Euphorbia hirta, Ficus racemosa, Mangifera indica, Mimosa pudica, Punica granatum, Rauvolfia serpentine, Shorea robusta,* and *Withania somnifera* to treat different human ailments in 9 (Aches and Fever, Breast Diseases, Dental and Oral System, Digestive System, Endocrine System, Integumentary System, Reproductive System, Skeletal System, and Urogenital System)and 6(Aches and Fever, Digestive System, Integumentary System, Liver System, Respiratory System, and Skeletal System) ailment categories/organ system categories respectively. It is observed that out of 13 plant species, 4 plant species, namely *Achyranthes aspera* (Digestive System), *Cissus quandrangularis* (Skeletal System), *Mangifera indica* (Digestive system), and *Rauvolfia serpentine* (Aches and Fever) are used by both of them to treat ailments in the same ailment category. It is observed that out of 13 Plant species, 1 plant species *Ficus racemosa* (Urogenital and Endocrine System) is used by the *Mandi* to treat ailments in more than one ailment categories/organ system categories. It is observed that out of 13 plant species, 6 plant species, namely *Achyranthes aspera* (Aches and Fever, Digestive System, and Integumentary System), *Andrographis paniculata* (Integumentary System and Digestive System), *Ficus racemosa* (Digestive System and Liver System), *Mangifera indica* (Digestive System and Integumentary System), *Punica granatum* (Digestive System and Liver System), *Rauvolfia serpentine* (Aches and Fever, Integumentary System and Skeletal System), and *Shorea robusta* (Digestive System and Integumentary System) are used by the *Kandhas* to treat ailments in more than one ailment categories/organ system categories.

It is observed that out of 13 plant species, the *Mandi* use 3 plants species, namely *Andrographis paniculata, Euphorbia hirta,* and *Rauvolfia serpentine* to treat communicable diseases such as cold, Malaria and Pyrexia; and 10 plant species, namely *Abutilon indicum, Achyranthes aspera, Asparagus racemosus, Cissus quadrangularis, Ficus racemosa, Mangifera indica, Mimosa pudica, Punica granatum, Shorea robusta, and Withania somnifera* to treat non-communicable diseases such as Azoospermia, Bone Fracture, Canker sore, Common Skin Diseases, Cystitis, Diabetics, Diarrhoea, Dysentery, Intestinal Worm Infestation, Mastitis in Women, Nociceptive Pain, Piles, and Pyrexia. It is observed that out of 13 plant species, the *Kandhas* use 2 plant species, namely *Achyranthes aspera* and *Euphorbia hirta* to treat communicable diseases such as Common Skin Diseases, Malaria and Pyrexia; and 11 plant species, namely *Abutilon indicum, Andrographis paniculata, Asparagus racemosus, Cissus quadrangularis, Ficus racemosa, Mangifera indica, Mimosa pudica, Punica granatum, Rauvolfia serpentine, Shorea robusta, and Withania somnifera* to treat non-communicable diseases such as Angioedema, Animal Bites, Asthma, Bilious Fever, Body Ache, Bone Fracture, Candidiasis, Cold, Dermatitis,

Table 9.2: Comparative Ethno-medico-botany of the *Mandi* in Bangladesh with the *Kandhas* in Kandhamal District of Orissa

Sl.No.	*Botanical Name*	*Part(s) Used*		*Biomedical Terms for Ailments*		*Organ System Categories*	
		Mandi	*Kandhas*	*Mandi*	*Kandhas*	*Mandi*	*Kandhas*
1	*Abutilon indicum*	Leaf and Root	Leaf	Mastitis in Women	Jaundice	Breast Diseases	Liver System
2	*Achyranthes aspera*	Root	Leaf	Interstitial Cystitis and Intestinal Worm Infestation	Bilious Fever, Fever, Rheumatic Pain, Typhoid and Wounds	Digestive System	Aches and Fever, Digestive System, and Integumentary System
3	*Andrographis paniculata*	Leaf	Leaf	Fever and Nociceptive Pain	Dermatitis, Intestinal Worm Infestation andItches	Aches and Fever	Digestive System and Integumentary System
4	*Asparagus racemosus*	Root	Root	Azoospermia	Dysentery	Reproductive System	Digestive System
5	*Cissus quadrangularis*	Whole Plant	Stem	Bone Fracture	Bone Fracture	Skeletal System	Skeletal System
6	*Euphorbia hirta*	Leaf	Root and Stem	Skin Diseases	Candidiasis, Cold, Intermittent Pyrexia, and Vitiligo	Integumentary System	Aches and Fever and Others
7	*Ficus racemosa*	Seed and Young Twig	Root	Diabetics and Piles	Diarrhoea, Hyperpyrexia, Jaundice, Indigestion, and Wounds	Endocrine System and Urogenital System	Digestive System and Liver System

Sl.No.	Botanical Name	Part(s) Used		Biomedical Terms for Ailments		Organ System Categories	
		Mandi	Kandhas	Mandi	Kandhas	Mandi	Kandhas
8	*Mangifera indica*	Stem Bark	Leaf and Stem Bark	Dysentry	Diarrhoea, Dysentery and Wounds	Digestive System	Digestive System and Integumentary System
9	*Mimosa pudica*	Root	Root	Mastitis in Women	Wounds	Breast Diseases	Integumentary System
10	*Punica granatum*	Young Twig and Root bark	Root and Young Fruit	Intestinal Worm Infestation	Diarrhoea, Dysentery and Jaundice	Digestive System	Digestive System and Liver System
11	*Rauvolfia serpentine*	Root and Leaf	Root	Malaria	Anasarca, Body Ache, and Rheumatism	Aches and Fever	Aches and Fever, Integumentary System and Skeletal System
12	*Shorea robusta*	Stem Bark	Bark	Canker Sore	Animal Bites, Diarrhoea and Dysentery	Dental and Oral System	Digestive System and Integumentary System
13	*Withania somnifera*	Root	Root	Diarrhoea	Asthma	Digestive System	Respiratory System

Diarrhoea, Dysentery, Hyperpyrexia, Indigestion, Intermittent Fever, Intestinal Worm Infestation, Itches, Jaundice, Rheumatism, Rheumatic Fever, Typhoid, Vitiligo, and Wounds.

3.2. Comparative Ethno-medico-botany of the *Mandi* in Bangladesh with the *Paliyans* of the Sirumalai Hills in Dindigul District of Tamil Nadu

Comparison between the *Mandi* and the *Paliyans* (Table 9.3) reveals that both of them use 6 plant species, namely *Achyranthes aspera, Aristolochia indica, Artemisia nilagirica, Asparagus racemosus, Centella asiatica, and Punica granatum* to treat different human ailments in 4 ailment categories/organ system categories (Digestive System, Integumentary System, Liver System, and Reproductive System) and 4 ailment categories/organ system categories (Dental and Oral System, Integumentary System, Respiratory System, and Urogenital System) respectively. It is observed that out of 6 plant species, 2 plant species, namely *Aristolochia indica* (Integumentary System), and *Centella asiatica* (Integumentary System) is used by both of them to treat ailments in the same ailment category. It is observed that out of 6 Plant species, *Artemisia nilagirica* (Integumentary system and Digestive System) is used by the *Mandi* to treat ailments in more than one ailment categories/organ system categories. It is observed that no plants are used by the *Paliyans* to treat ailments in more than one ailment categories/organ system categories.

It is observed that out of 6 plant species, the *Mandi* use *Artimisia nilagrica* to treat a communicable disease, Leprosy; and 5 plant species *Achyranthes aspera, Aristolochia indica, Asparagus racemosus, Centella asiatica, and Punica granatum*to treat non-communicable diseases such as Animal Bites, Azoospermia, Cystitis, Dysentery, Intestinal Worm Infestation, Jaundice, Hepatospleenomegaly/Liver Pain, Rheumatism, and Wounds. It is observed that out of 6 plant species, the *Paliyans* use *Aristolochia indica* to treat communicable Common Skin Diseases; and 5 plant species, namely *Achyranthes aspera, Artimisia nilagrica, Asparagus racemosus, Centella asiatica*, and *Punica granatum*to treat non-communicable diseases such as Animal Bites, Asthma, Urolithiasis, Oral Problems, and Wounds.

3.3. Comparative Ethno-medico-botany of the *Mandi* in Bangladesh with the *Pawras* of the Satpura Hills, Maharashtra

Comparison between the *Mandi* and the *Pawra* (Table 9.4) reveals that both of them use 1 plant species *Asparagus racemosa* to treat 1 human ailment in different ailment categories/organ system categories (Dental and Oral System and Reproductive System). It is observed that *Asparagus racemosa* is not used by the *Mandi* to treat any communicable diseases but used to treat a non-communicable reproductive system disorder, Sexual Dysfunction. It is observed that *Asparagus racemosa* is not used by the *Pawras* to treat any communicable diseases but is used to treat a non-communicable Oral Problem, Gingivitis.

Table 9.3: Comparative Ethno-medico-botany of the *Mandi* in Bangladesh with the *Paliyans* of the Sirumalai Hills in Dindigul District of Tamil Nadu

Sl.No.	*Botanical Name*	*Part (s) Used*		*Biomedical Terms for Ailments*		*Organ System Categories*	
		Mandi	*Paliyans*	*Mandi*	*Paliyans*	*Mandi*	*Paliyans*
1	*Achyranthes aspera*	Root	Root	Cystitis and Intestinal Worm Infestation	Animal Bites	Digestive System	Integumentary System
2	*Aristolochia indica*	Leaf and Stem	Leaf	Wounds	Skin Diseases	Integumentary System	Integumentary System
3	*Artemisia nilagirica*	Leaf	Leaf	Dysentery, Leprosy, and Liver Pain	Asthma	Digestive System, Integumentry System, and Liver System	Respirational Sytem
4	*Asparagus racemosus*	Root	Tubers	Azoospermia	Nephrolithiasis	Reproductive System	Urogenital System
5	*Centella asiatica*	Leaf and Seed	Leaf	Animal Bites, Jaundice and Rheumatism	Wounds	Integumentary System and Liver System	Integumentary System
6	*Punica granatum*	Young Twig and Root Bark	Fruit	Intestinal Worm Infestation	Oral Problems	Digestive System	Dental and Oral System

3.4. Comparative Ethno-medico-botany of *the Mandi* in Bangladesh with the Sub-communities: Tolchha and Marcha of the *Bhotiya* in the Central Himalaya

Comparison between the *Mandi* and (2 Sub-communities of) the *Bhotiya* (Table 9.5) reveals that both of them use 2 plant species (*Achyranthes aspera* and *Centella asiatica*) to treat different human ailments in 4 (Digestive System, Integumentary Problems, Liver Problems, and Skeletal Problems) and 1(Digestive System) ailment categories/organ system categories. It is observed that both of them use 1 plant species*Achyranthes aspera* to treat ailments in the same category (Digestive System). It is observed that out of 2 plant species (*Achyranthes aspera* and *Centella asiatica*), 1 plant species *Centella asiatica* is used by the *Mandi* to treat ailments in more than one ailment categories/organ system categories (Skeletal System, Liver System, and Integumentary System). It is observed that no plants are used by the sub-communities of *Bhotiya* to treat ailments in more than one ailment categories/organ system categories.

It is observed that out of 2 plant species, the *Mandi* use no plant species to treat communicable diseases and use 2 plant species, namely *Achyranthes aspera* and *Centella asiatica*to treat non-communicable diseases such as Intestinal Worm Infestation, Cystitis, Rheumatism, Jaundice, and Animal Bites. It is observed that out of 2 plant species, the sub-communities of *Bhotiya* use no plant species to treat communicable diseases and use 2 plant species, namely *Achyranthes aspera* and *Centella asiatica* to treat a non-communicable disease Intestinal Worm Infestation.

3.5. Communicable and Non-Communicable Disease

Tables 9.6–9.10 shows the number of plants used for communicable and non-communicable disease by the *Mandi*, *Kandha*, *Pawra*, *Paliyans,* and *Bhotiya* tribes. It represents that 7 plants namely (*Andrographis paniculata* : Fever, *Artimisia nilagirica*: Leprosy, *Euphorbia hirta*: Cold, *Rauvolfia serpentine*: Malaria, *Achyranthes aspera*: Typhoid, *Euphorbia hirta*: Cold, *Aristalochia indica*: Leprosy) were used by these 5 tribes to cure communicable disease and 13 plants namely (*Abutilon indicum*: Jaundice *Andrographis paniculata*: Dermatitis and Intestinal Worm Infestation and Itches *Asparagus racemosus*: Dysentry, Animal Bites, Sexual Dysfunction, Gingivitis and Nephrolithiasis, *Cissus quadrangularis*: Bone Fracture, *Ficus racemosa*: Hyperpyrexia, Jaundice, Wounds, Diarrhoea and Indigestion, *Mangifera indica*: Diarrhoea, Dysentery, and Wounds, *Mimosa pudica*: Wounds, *Punica granatum*: Jaundice, Diarrhoea, Oral Problems and Dysentery, *Rauvolfia serpentine*: Body Ache, Anasarca and Rheumatism, *Shorea robusta*: Diarrhoea, Dysentery, An Animal Bite, *Withania somnifera*: Asthma, *Artemisia nilagirica*: Asthma and *Centella asiatica*: Wounds) were used to treat non-communicable diseases. It has been found that out of these 20 plants, 2 plants such as *Asparagus racemosus* (5 uses) and *Punica Granatum* (4 uses) have multiple uses and are mostly preferred by these 5 tribes.

Table 9.4: Comparative Ethno-medico-botany of the *Mandi* in Bangladesh with the *Pawras* of the Satpura Hills, Maharashtra

Sl.No.	*Botanical Name*	*Part (s) Used*		*Biomedical Terms for Ailments*		*Organ System Categories*	
		Mandi	*Pawras*	*Mandi*	*Pawras*	*Mandi*	*Pawras*
1.	*Asparagus racemosus*	Root	Root	Sexual Dysfunction	Gingivitis	Reproductive System	Oral and Dental System

Table 9.5: Comparative Ethno-medico-botany of *the Mandi* in Bangladesh with the Sub-communities: Tolchha and Marcha of the *Bhotiya* in the Central Himalaya

Sl.No.	*Botanical Name*	*Part (s) Used*		*Biomedical Terms for Ailments*		*Organ System Categories*	
		Mandi	*Bhotiya*	*Mandi*	*Bhotiya*	*Mandi*	*Bhotiya*
1	*Achyranthes aspera*	Root	Root	Interstitial Cystitis and Intestinal Worm Infestation	Intestinal Worm Infestation	Digestive System	Digestive System
2	*Centella asiatica*	Leaf	Whole Plant	Animal Bites, Jaundice Nociceptive Pain, and Rheumatism	Intestinal Worm Infestation	Integumentary System, Liver System, and Skeletal System	Digestive System

Table 9.6: List of Plants Used by the *Mandi* to Treat Communicable and *Non-communicable Diseases*

Communicable Diseases	*Non-communicable Diseases*
☆ *Andrographis paniculata* : Fever ☆ *Artimisia nilagirica*: Leprosy ☆ *Euphorbia hirta* : Cold ☆ *Rauvolfia serpentina* : Malaria	☆ *Abutilon indicum*: Mastitis in Women ☆ *Achyranthes aspera*: Intestinal Worm ☆ Infestation and Interstitial Cystitis ☆ *Aristolochia indica*: Wounds ☆ *Asparagus racemosus*: Azoospermia ☆ *Centella asiatica*: Rheumatism, Animal Bites and Jaundice ☆ *Cissus quadrangularis*: Bone Fracture ☆ *Ficus racemosa*: Piles and Diabetics ☆ *Mangifera indica*: Dysentery ☆ *Mimosa pudica*: Wounds ☆ *Punica granatum*: Jaundice, Diarrhoea and Dysentery ☆ *Shorea robusta*: Canker sore ☆ *Withania somnifera*: Diarrhorea

Table 9.7: List of Plants Used by the *Kandhas* to Treat Communicable and Non-communicable Diseases

Communicable Diseases	*Non-communicable Diseases*
☆ *Achyranthes aspera*: Typhoid ☆ *Euphorbia hirta*: Cold	☆ *Abutilon indicum*: Jaundice ☆ *Andrographis paniculata*: Dermatitis and Intestinal Worm Infestation and Itch*es* ☆ *Asparagus racemosus*: Dysentry ☆ *Cissus quadrangularis* : Bone Fracture ☆ *Ficus racemosa*: Hyperpyrexia, Jaundice, Wounds, Diarrhoea and Indigestion ☆ *Mangifera indica*: Diarrhoea, Dysentery and Wounds ☆ *Mimosa pudica*: Wounds ☆ *Punica granatum*: Jaundice, Diarrhoea and Dysentery ☆ *Rauvolfia serpentine*: Body Ache, Anasarca and Rheumatism ☆ *Shorea robusta*: Diarrhoea, Dysentery, Animal Bite ☆ *Withania somnifera*: Asthma

Table 9.8: List of Plants Used by the *Paliyans* to Treat Communicable and Non-communicable Diseases

Communicable Diseases	*Non-communicable Diseases*
✰ *Aristalochia indica*: Leprosy	✰ *Achyranthes aspera*: Animal Bites ✰ *Artemisia nilagirica*: Asthma ✰ *Asparagus racemosus*: Nephrolithiasis ✰ *Centella asiatica*: Wounds ✰ *Punica granatum*: Oral Problems

Table 9.9: List of Plants Used by the *Pawra* to Treat Communicable and Non-communicable Diseases

Communicable Diseases	Non-communicable Diseases
	✰ *Asparagus racemosus* :Sexual Dysfunction and Gingivitis

Table 9.10: List of Plants Used by the *Bhotiya* to Treat Communicable and Non-communicable Diseases

Communicable Diseases	*Non-communicable Diseases*
	✰ *Achyranthes aspera*: Intestinal Worm Infestation ✰ *Centella asiatica*: Intestinal Worm Infestation

4 Conclusion

There is a well-developed ethnomedicinal system present among the *Mandi, Kandhas, Paliyans, Pawra,* and *Bhotiya* (indigenous people) in their settlements where ethno-medicinal surveys were carried out by the researchers. The indigenous people have great expertise with the plants of their surroundings. The ethno-medicinal information gathered from the comparative ethno- medico botany of the *Mandi, Kandhas, Paliyans, Pawra,* and *Bhotiya* is useful for further research in the field of pharmacology. The information could also be useful for the pharmacologists, physicians, phytochemists, botanists, and alike interested in the development of alternative therapies and pharmaceuticals. For lesser-known plant species, such a secret treasure of ethno-medicinal information could prove beneficial in phyto-pharmacological research for the discovery of new therapeutic drugs. The present study could be of help to develop new drugs based on traditional medicine for the effective remedy of different human ailments. The ethno-medicinal information generated from the present study regarding the traditional medicinal plants used by

Table 9.11: Comparing the Mandi Community in Banglades with the Other Communities of India Using Jaccard Index

Tribal Settlements	*References*	*Year*	*No. of Plant Species Reported in the Focal Research Paper (A)*	*No. of Plant Species Reported in the another Research Paper Chosen for Comparison with Focal Research Paper (B)*	*Same Plant Species Reported in the Focal Research Paper and Other Research Paper Chosen for Comparison (C)*	*Jaccard Index*
Khadas in Kandhamal District of Orissa	Soumit *et al.*	2006	109	98	13	6.7015
Paliyans of Sirumalai Hills in Dindigul Districtof Tamil Nadu	Karuppusamy	2007	109	90	6	3.1086
Pawra of Satpura Hills, Maharashtra	Jagtap *et al.*	2009	109	79	1	0.535
Bhotiya Community in Central Himalaya	Phondani *et al.*	2010	109	86	2	1.0362

the *Mandi, Kandhas, Paliyans, Pawra,* and *Bhotiya* needs a thorough phytochemical investigation including alkaloid extraction and isolation along with a few clinical trials in the future course of time. This could help in creating mass awareness regarding the need for conservation of such valuable medicinal plants and also in the promotion of ethno- medico-botanical knowledge within the region, Given seeing the exploitation of biodiversity, traditional knowledge in developing countries like India is eroding at a faster rate. Therefore, it is felt as an urgent need to inventories and record all ethno-medicinal information available in diverse ethnic communities before the traditional healthcare practice is completely lost.

5. Recommendations

Based on the results of the present study, the following recommendations are suggested:

- ☆ Resources, especially plant resources are integral to the life of all biota, as they are the primary food producers. The life of world biota is directly or indirectly dependent on plant resources. Thus, the *Mandi, Kandhas, Paliyans, Pawra,* and *Bhotiya* (indigenous people) in their settlements should be involved in conservation and management plans of plant resources or their indigenous knowledge in their locality.
- ☆ Traditional medicinal plants are central to the indigenous cultures and material needs. Therefore, formal and non-formal education systems should be designed to create a positive attitude among the young by integrating into the curricula about the traditional uses of plants in general and medicinal plants in particular
- ☆ Ethno-medicinal studies should be promoted as they are significant for the discovery of new crude drugs from the reported medicinal plants. Government and particularly NGOs should involve in ethno-medicinal and ethnopharmaceutical studies to compile and preserve such crucial indigenous knowledge.

6. References

1. Aziz MA, Adnan M, Khan AH, Sufyan M, Khan SN. 2018; Cross-cultural analysis of medicinal plants commonly used in ethnoveterinary practices at South Waziristan Agency and Bajaur Agency, Federally Administrated Tribal Areas (FATA), Pakistan. *J Ethnopharmacol.* **210**: 443– 468.
2. Bannerman, R. H. 1982. Traditional Medicine in Modern Health Care, World Health Forum. **3**: 8-13.
3. Bedoya, L.M., P. Bermejo and M.J. Abad, 2009. Anti-infectious activity in the Cistaceae family in the Iberial Peninsula. *Mini Reviews in Medicinal Chemistry,* **9(5)**: 519-525.
4. Behera, S.K., Panda, A., Behera, S.K. and Misra, M.K., 2006. Medicinal plants used by the Kandhas of Kandhamal district of Orissa. *IJTK.* **5(4)**: 519-528

5. Chandra Prakash Kala, 2005. Ethnomedicinal botany of the Apatani in the Eastern Himalayan region of India Journal of Ethnobiology and Ethnomedicine **1**: 11
6. Cotton, C. M. 1996. Ethnobotany: Principles and Applications. Chichester, UK: John Willey and Sons Ltd. **424**.
7. Dutta BK, Dutta PK, 2005. Potential of ethnobotanical studies in North East India: an overview. *Indian Journal of Traditional Knowledge* **4(1)**: 7-14
8. Fransworth, N. R. 1988. Screening Plants for New Medicines. In: Wilson, E. O. (Ed.) Biodiversity. Washington DC: National Academy Press. pp. 83-97.
9. Fransworth, N. R. 1994. Ethnopharmacology and Drug Development. Cipa Found Symp. **185**: 42-51.
10. Gairola S, Sharma J, Bedi YS. 2014; A cross-cultural analysis of Jammu, Kashmir, and Ladakh (India) medicinal plant use. J Ethnopharmacol. **155**: 925–986.
11. Govaerts R, How many species of seed plants are there? Taxonomy 2001, **50:** 1085-1090
12. Heinrich M, Ankil A, Frei B, Weimann C, Sticher O, 1998. Medicinal plants in Mexico, healer's consensus, and cultural importance. *Soc Sci Med.* **47**: 1859–1871.
13. https: //en.wikipedia.org/wiki/Bhotiya
14. https: //www.ethnologue.com/language/grt
15. Ignacimuthu, S., Ayyanar, M. and Sankara Sivaraman, K. 2006. Ethnobotanical Investigations among Tribes in Madurai District of Tamil Nadu (India). *J. Ethnobiol. Ethnomed.* **2(25)**: 2-7.
16. Jagtap, S. D., Deokule, S. S.1, Pawar, P. K. and Harsulkar, A. M. 2009. Traditional Ethnomedicinal Knowledge Confined to the Pawra Tribe of Satpura Hills, Maharashtra, India *Ethnobotanical Leaflets.* **13**: 98-115.
17. Jain, S. K., and Mudgal, V. 1999. A Handbook of Ethnobotany. Bishen Singh Mahendra Pal Singh Publications, Dehra Dun, India.
18. Kala, C. P. 2005. Current Status of Medicinal Plants Used by Traditional Vaidyas in Uttaranchal State of India. *Ethnobot. Res. Appl.* **3**: 267-278.
19. Karuppasamy, S. 2007. Medicinal Plants Used by Paliyan Tribes of Sirumalai Hills of Southern India. *Nat. Prod. Radiance.* **6:** 436-442.
20. Kumar, P., 2014. Ethno medicinal plants used for oral health care in India. *Justicia*, **6:** 7.
21. Lardos A, Heinrich M. 2013; Continuity and change in medicinal plant use: the example of monasteries on Cyprus and historical atrosophia texts. *J.Ethnopharmacol.* **150**: 202–214.

22. Martin, G. J. 1995. Ethnobotany: A Methods Manual. Chapman and Hall, London. pp. 265-270.

23. Mukul, S. A., Uddin, M. B. and Tito, M. R. 2007. Medicinal Plant Diversity and Local Healthcare among the People Living in and around a Conservation Area of Northern Bangladesh. *IJFUM*. **8**: 50-63.

24. Nitesh Kumar, Ashwani Kumar Jakhar, Rajaram Choyal.2014. Traditional Uses Of Some Medicinal Plants Of Hamirpur District Of Himachal Pradesh For The Treatment Of Diabetes *Inter. J. Adv Res*, **2(2)**: 131-138

25. Nithya, N. R. 2014. Globalization and the Plight of Tribals: The Case of Kerala, India. *Dawn Journal*. **3**. 727-758.

26. Park, E. J. and Pezzutto, J. M. 2002. Botanicals in Cancer Chemoprevention. *Cancer Metastasis Rev*. **21**: 231-255.

27. Partha, P., and Hossain, A. 2007. Ethnobotanical investigation into the Mandi ethnic community in Bangladesh. *Bangladesh Journal of Plant Taxonomy*, ***14*(2)**: 129-145.

28. Pei, S. J. 2001. Ethnobotanical approaches of traditional medicinal studies: Some experiences from Asia, *Pharmaceut Biol*, **39**: 74-79

29. Phondani, P.C., Maikhuri, R.K., Rawat, L.S., Farooquee, N.A., Kala, C.P., Vishvakarma, S.R., Rao, K.S. and Saxena, K.G., 2010. Ethnobotanical uses of plants among the Bhotiya tribal communities of Niti Valley in Central Himalaya, India.

30. Pieroni A, Giusti ME, Quave C. 2011 Cross-cultural ethnobiology in the western balkans: medical ethnobotany and ethnozoology among albanians and serbs in the Pešter Plateau, Sandžak, South-Western Serbia. *Hum Ecol*. **39**: 333–349.

31. Prasad, R. R. *Encyclopaedic profile of Indian tribes*. Vol. 1. Discovery Publishing House, 1996.

32. Rahmatullah, M., Azam, M.N.K., Rahman, M.M., Seraj, S., Mahal, M.J., Mou, S.M., Nasrin, D., Khatun, Z., Islam, F. and Chowdhury, M.H., 2011. A survey of medicinal plants used by Garo and non-Garo traditional medicinal practitioners in two villages of Tangail district, Bangladesh. *American Eurasian Journal of Sustainable Agriculture*, **5**: 350-357.

33. Rajiv, S. K. 1998. Ethnobotany: The Renaissance of Traditional Herbal Medicine, INA Publishers, Jaipur.

34. Ramakrishnan PS: 1992. Tropical forests, exploitation, conservation, and management. Impact of Science on Society, **42(166)**: 149-162.

35. Rao, 2015. Role of cross cultural ethno-botany in search of newer drug plants: Some concerns and issues. 3rd International Conference and Exhibition on Pharmacognosy, Phytochemistry and Natural Products. October 26-28, 2015 Hyderabad, India

36. Rishton, G.M., 2008. Natural products as a robust source of new drugs and drug leads: past successes and present day issues. *American Journal of Cardiology*, **101(10A)**: 43D–49D.

37. Salminen, A., M. Lehtonen, T. Suuronen, K. Kaarniranta and J. Huuskonen, 2008. Terpenoids: natural inhibitors of NF-kB signaling with anti-inflammatory and anticancer potential. *Cellular and Molecular Life Sciences*, **65(19)**: 2979-2999.

38. Sanjappa, M.Plant Diversity in India- Status, Conservation, and Challenges, In: Proceedings of the XXVIII conference of Indian Botanical Society, October 24-26, 2005. BSI, Dehradun, pp.5-6

39. Sanjeet Kumar, P. K. Jena, S. Sabnam, M. Kumari and P. K. Tripathy 2012, "Study of plants used against the skin diseases with special reference to Cassia fistula L. among the king (Dongaria Kandha) of Niyamgiri: A primitive tribe of Odisha, India", *Int. J. Drug Dev. and Res.*, **4(2)**: 256-264

40. Savnur, H. C. 1993. A Handbook of Ayurvedic Materia Medica, VI, Dr. Jarthar and Sons, Maruthi Street, Belgaum, Karnataka, India. pp. 258-261.

41. Schmidt, B., D.M. Ribnicky, A. Poulev, S. Logendra, W.T. Cefalu and I. Raskin, 2008. A natural history of botanical therapeutics. *Metabolism*, **57 (1)**: S3-9.

42. Shaikh, B.T. and J. Hatcher, 2005. Complementary and Alternative Medicine in Pakistan: Prospects and Limitations. *Evidence-based Complementary and Alternative Medicine*, **2(2)**: 139- 142.

43. Shenbaham, K. 2016. Culture and Lifestyle of Palani Tribes at Palani Hill, Dindigul District. Int. *J. Human. Soc. Sci. Res.* **2(3)**: 37-40.

44. Shroff, J. 1997. Forest Policy, and Tribal Development. In: Forest Management in Tribal Areas (Eds. Mohapatra, P. M. and Mohapatra, P.) Concept Publishing Company, Mohan Garden, New Delhi, India.

45. Silva, T. D. 1997. Industrial Utilization of Medicinal Plants in Developing Countries. In: Medicinal Plants for Forest Conservation and Health Care. (Eds. Bodeker, G., Bhat, K. K. S. Burley, J. and Vantomme, P.) FAO Non-wood Forest Products Series No. 11. FAO, Rome, Italy. pp. 34-44.

46. Singh, M., Walia, S. and Singh, A. K. 2013. Phytochemical Characterization and Antibacterial Activity of Lindenbergia indica Vatke. A Common Wall Flora against Some Human Pathogens in Doon Valley, Uttarakhand. *Indian J. Nat. Prod. Resour.* **4**: 138-145.

47. Srithi, K.; Balslev, H.; Wangpakapattanawong, P.; Srisanga, P.; Trisonthi, C. Medicinal plant knowledge and its erosion among the Mien (Yao) in northern Thailand. *J. Ethnopharmacol.* 2009, **123**: 335–342.

48. Tang, S. Y. and Haliwel, B. 2010. Medicinal Plants and Antioxidants: What do We Learn from Cell Culture and Caenorhabditiselegans Studies *Biochem. Biophys. Res. Commun. Mini Review*. **394**: 1-5.

49. Tapsell, L.C., I. Hemphill, L. Cobiac, C.S. Patch, D.R. Sullivan, M. Fenech, S. Roodenrys, J.B. Keogh, P.M. Clifton, P.G. Williams, V.A. Fazio and K.E. Inge, 2006. Health benefits of herbs and spices: the past, the present, the future. *Medical Journal of Australia*, **185(4)**: S4–24.

50. WHO survey. In medicinal plants (Eds. Haq. I.), (Hamdard Foundation Press, Karachi), 1993, **13**.

Chapter 10

Comparative Ethno-Medico-Botany of the *Paliyans* in the Sadhuragiri Hills, Virudhunagar District with the *Paliyans* in Some Other Areas of Tamil Nadu

N. Ahamed Kabir[1], *K. Afrin Banu*[2] *and A. Doss*[3]

[1]*Research Scholar, PG and Research Department of Botany, V.O. Chidambaram College, Thoothukudi*

[2]*PG and Research Department of Botany, V. O. Chidambaram College, Thoothukudi*

ABSTRACT

In the present study the traditional medicinal plants used for treating various human ailments by the Paliyans colonized in the Sadhuragiri Hills in the Southern Western Ghats of Srivilliputhur Taluk, Virudhunagar District, Tamil Nadu, India are compared with the traditional medicinal plants used for treating various human ailments by the Paliyans populated in and around the Madurai District Forest Areas; in the Sirumalai Hills range of the Western Ghats in Dindigul District; in the Shenbagathope Forest Areas found on the Eastern slopes of the Western Ghats in Srivilliputhur Taluk, Virudhunagar District and in some villages of Virudhunagar District to know about the uniformity and variation in ailments treated, part(s) used, use of fresh, dry plant part(s), use of a single or a combination of plants for remedy preparation, mode of remedy preparation, dose requirement and duration. The ethno-medicinal information of 239 plant species as having medicinal properties against about 69 human ailments under 22 categories was gathered from 5 research papers chosen for this comparative study. Since the Paliyans settlements have good ethno-botanical potential for traditional medicinal plants, they could be suitable places for further comparative ethno-medico-botany studies and ethno-pharmaceutical studies. This first comparative ethno-medico-botany of the Paliyans settled in some areas of Tamil Nadu with the Paliyans settled in some other areas of Tamil Nadu is not yet followed by investigators in the field of ethnobotany.

***Keywords**: Ethno-medico-botany, Paliyans*

1. Introduction

1.1. Indigenous Knowledge

The indispensable dependency of humanity upon plants for their livelihood was primarily started by domestication and dates back 10,000 years (Martin, 1995). Indigenous knowledge on plants appeared when humans started and learned how to use plants (Posey, 1999). Indigenous knowledge suggests to the accumulation of knowledge, rule, standards, skills and mental sets, which are possessed by local people in a particular Areas (Quanash, 1998). Indigenous knowledge on remedies in many countries passes from one generation to the other orally with great secrecy (Jansen, 1981). Such secrete and crude transfers make indigenous knowledge or ethnomedicinal knowledge vulnerable to distortion and in most cases, some of the lore is lost at each point of transfer (Getahun, 1976), hence there is a need for systematic documentation of such useful knowledge through ethnobotanical research.

1.2. Traditional Medicine

The development of traditional medicinal systems incorporating plants as means of therapy can be traced back to the Middle Paleolithic age some 60,000 years ago as found from fossil studies (Solecki and Shanidar, 1975). World Health Organization has stated that 80 per cent of the world's population depends on traditional medicine for its primary healthcare and has become indispensable for its survival (Hiremath and Taranath, 2013). Traditional medicine is widely used and accounts for about 40 per cent of all healthcare delivered (WHO, 2002-2005). The use of traditional medicine is widespread and plants still present a large source of novel biological active compounds with different activities, including anti-inflammatory, anti-cancer, anti-viral, anti-bacterial and cardio protective activities (Yan *et al.,* 2002).

According to Kibebew (2001), about 75-90 per cent of the rural population in the world (excluding western countries) relies on traditional medicine as system. This is not only because of poverty that people cannot afford to buy expensive modern drugs, but traditional systems are also more culturally acceptable and meet the psychological needs in a way modern medicine does not. The plant based traditional knowledge has become a recognized tool in search for new sources of drugs (Bruce and Meeus, 2002). Traditional knowledge of medicinal plants has always guided the search for new cures. In spite of the advent of modern high thoroughput drug discovery and screening techniques, traditional knowledge systems have given clues to the discovery of valuable drugs (Buenz*et al.*, 2004).

1.3. Tribal People in India

Most ethnobotanical studies have been restricted to studies of tribal and rural people to record their knowledge and use of plants and to search for new sources of herbal drugs, edible plants and other properties of plants of value to man (Jain and Mudgal, 1999). From the study of the anthropological literature, it appears that the term tribe has been used for those groups of human beings, whose place of residence is situated in remote Areass like hills, forest, sea coasts and islands; and whose style of life is quite different from the present day civilized men. The forest occupies a central position in tribal culture and economy. The tribal way of life is very much dictated by the forest right from birth to death. India is also characterized by having second largest tribal (*Adivasis*) population in the world. It is ironical that the poorest people of India are living in the Areass of richest natural resources. The Imperial Gazetteer of India, 1911, deûnes a tribe as a "collection of families bearing a common name, speaking a common dialect, occupying or professing to occupy a common territory and is not usually endogamous though originally it might have been so." According the 2011 census, tribals constitute 8.2 per cent of the total population of the country. The tribal people of India are called "Scheduled Tribes" in the Indian Constitution. In India, 461 ethnic groups are recognized as Scheduled Tribes, and these are considered to be India's indigenous peoples. In mainland India, the Scheduled Tribes are usually referred to as *Adivasis,* which literally means indigenous peoples. According to the Anthropological Survey, a total of 4,635 communities are now to be found in India. Out of this total, "tribal" or Indigenous communities number 732. The largest concentrations of indigenous peoples are found in the seven states of north-east India, and the so-called "central tribal belt" stretching from Rajasthan to West Bengal. More than half of the Scheduled Tribes population is concentrated in the States of Madhya Pradesh, Chhattisgarh, Maharashtra, Odisha, Jharkhand and Gujarat. There are over 700 Scheduled Tribes notiûed under Article 342 of the Constitution of India, spread over different states and union territories of the country. Many tribes are present in more than one state. The largest numbers of Scheduled Tribes (*i.e.* 62) are in the state of Orissa (Nithya, 2014).

Indian subcontinent is being inhabited by over 53.8 million tribal people in 5000 forest dominated villages of tribal community and comprising 15 per cent of the total geographical Areas of Indian landmasses, representing one of the greatest emporia of ethnobotanical wealth (Sajem and Gosai, 2006). Almost every section of Indian population uses plants as medicine and altogether about 7500 species of plants are being used by several ethnic communities. Particularly, tribal people collect and preserve locally available wild and cultivated plant species and practice herbal medicine to treat a variety of diseases and disorders. With enormously diversiûed ethnic groups and rich biological resources, India represents one of the great emporia of ethnobotanical wealth (Kala, 2005).

1.4. Ethnobotany

Harshberger in 1895 coined the term ethnobotany to indicate plants used by the aboriginals. It included the study and evaluation of plant-human relations in all phases and the effect of plant environment on human society. Subsequently Schultes (1962) defined, ethnobotany as "the study of the relationship which exists between people of primitive societies and their plant environment". Ethnobotany is a multidisciplinary subject deals with the total relationships between plants and people in context to their culture. It is diûcult to state exactly when the term ethnobotany became part of modern science. However, it can be traced back to the time when humans started making conscious interaction with plants and animals.

Ethnobotanical work seems to have started with (Christopher Columbus in 1492), at a time when he brought tobacco, maize, spices and other useful plants to Europe from Cuba (Cotton, 1996) and when other immigrants from the new world documented food, medicine and other useful plants of the Aztec, Maya and Inca peoples (Martin, 1995). The curative properties of some herbs were documented in Rigveda, Atharvaveda, Upanishads, Mahabharata and Puranas; written between 4500-1600 BC (Jain, 1994). It was followed by Atharvaveda, wherein the magico-religious aspects and utility of medicinal herbs were recorded. Later, eight division of Ayurveda, written probably between 2500 and 900 BC were the real foundation of ancient medicine. Detailed accounts of medicinal plants are also found in Chikitsa Sthanam of Sushruta Samhita, written between 600 BC and 300 AD (Bishagranta 1907-1916; Jain 1996; Jain and Mudgal, 1999). Over the past years, people who study the relationships between humans and plants in any ûeld (for example, medicine, food and forestry), and at any level (local, regional and so on) recognize that the approach to ethnobotanical studies has gone through conceptual and methodological changes since the classical works, starting from the coining of the name Ethnobotany by Harshberger in 1896 (Carrió and Vallès, 2012).

1.5. Importance of Ethnobotany

As stated by Martin (1995) to achieve more detailed and reliable information of plants and plant use, ethnobotanical study needs involvement of specialists from various disciplines, such as plant taxonomists, plant ecologists, anthropologists, linguists, economic botanists, pharmacologists and others. With such interdisciplinary and multidisciplinary approaches, ethnobotany is aimed at gathering and documenting indigenous botanical knowledge, cultural practice, use and management of botanical resources.

1.6. Ethnomedicine

Many tribal groups have been using several plant or animal products for medicinal preparations and these medicines are known as ethnomedicines (Pushpangadan and Atal, 1984). Tribals have deep belief in their native folklore medicine for remedies and they rely exclusively on their own herbal cure (Sajem and Gosai, 2006). In developing countries, there is an increasing attempt to incorporate traditional medicines, especially herbal preparations in the local healthcare systems and many modern researchers involved today to explore the huge potential of ethnobotanical knowledge for treating various diseases (Dutta and Dutta, 2005; Kala, 2005; Jain *et al.*, 2010; Jeyaprakash, 2011). However, the ethnomedicinal plants are under threat due to deforestation, overgrazing and their reckless utilization. So, it indicates the urgent need of their conservation. Conservation of biological resources as well as their sustainable use is important in preservation of traditional knowledge (UNU-IAS, 2013).

2. Materials and Methods

In the present study the traditional medicinal plants used for treating various human ailments by the *Paliyans* colonized in the Sadhuragiri Hills in the Southern Western Ghats of Srivilliputhur Taluk, Virudhunagar District, Tamil Nadu, India are compared with the traditional medicinal plants used for treating various human ailments by the *Paliyans* populated in and around the Madurai District Forest Areas; in the Sirumalai Hills range of the Western Ghats in Dindigul District; in the Shenbagathope Forest Areas found on the Eastern slopes of the Western Ghats in Srivilliputhur Taluk, Virudhunagar District and in some villages of Virudhunagar District to know about the uniformity and variation in ailments treated, part(s) used, use of fresh, dry plant part(s), use of a single or a combination of plants for remedy preparation, mode of remedy preparation, dose requirement and duration.

2.1. *Paliyans*

It is believed that the *Paliyans* are indigenous people of Palani Hills. In the Palani Hills they are found at an altitude of upto 2200 m. Generally the *Paliyans* are illiterate and they speak Tamil (mother tongue of Tamil Nadu). They are traditional Nomadic hunter-gatherers, honey hunters and foragers. Yams are their major food source. In the early part of the 20th century the *Paliyans* dressed scantily and lived in rock crevices and caves. Most have now have transformed to traders of forest products, food cultivators and beekeepers. (Shenbaham, 2016). The *Paliyans*, when compared to various tribal communities in Tamil Nadu constitute relatively a small group (Ignacimuthu *et al.*, 2006). The *Paliyans* are also found in the hilly regions of Madurai, Dindigul, Theni, Tirunelveli and Virudhunagar Districts. Reportedly the descendants of *Paliyans* still live in Kukul cave in Kodaikanal set high in the Palani

Hills at 2100 m. *Paliyans* used to live in small caves or in hut-like structure between two rocks. Still now some *Paliyans* live in rocky caves while roaming in interior forest. Some who are living in the interior forest build huts with the minimum of wood and grass (Shenbaham, 2016).

2.2. Physical Characteristics of *Paliyans*

Thurston, (1909) quoting from the writing of Rev. F. Dahman (1908) describes *Paliyans* as nomadic tribe, who for the most part rove in small parties through the jungle clad gorges (Sholas) that fringe the upper Palni plateau. Gardner describes *Paliyans*as moderately dark in complexion, short in stature, and their physical characteristics fall within the range of South and South East Asian (Gardner, 1972; Thomas *et al.,* 2003). Gardner (1972) describes the *Paliyans* physical characteristics to fall within the range of South and South East Asian-Australoid types formally referred as Negrito. Physically they are similar to the *Semong* of Malaya and other Indian tribal communities (Dahmen, 1908).

Gardner, the foremost English *Paliyan* Author, states that *Paliyans* have distinctive ways:

- ☆ To learn subsistence skills with minimal formal verbal instruction.
- ☆ To ascertain themselves what others denote by terms - in the face of substantial interpersonal knowledge variation.
- ☆ To "know" by testing empirically the validity of theories (*e.g.*, on the condition of a hunted animal (Lee *et al.,* 1999).

Gardner (1993) has described traditional hunting and gathering techniques in detail: foraging for yams and sago, hunting of small game, fishing, and honey collecting. He has shown that *Paliyan* data refute the century-old theory (revived by Bose in 1956 and by Fox in 1969) that India foragers are "professional primitives," unable to subsist on hunting and gathering alone. Gardner (2006) summarizes: "The *Paliyan* prescription for a good life - residing in a natural setting, cultivating a mature and independent self, respecting all others and fostering a peaceful society".

2.3. Categories of *Paliyans*

The *Paliyans* can be grouped into three categories based on their life styles, namely, nomadic, semi nomadic and settled. Nomadic *Paliyans* don't build houses. They live temporarily in rock caves called '*Pudai*'. Semi nomadic *Paliyans* build temporary houses and confine themselves to small territories. Most of their huts are dark with no window or any other opening to admit air. Settled *Paliyans* are more or less urbanized and live as agricultural laborers. Importance of traditional and folk medicine in the treatment of various human ailments is well recognized amongst these people (Ignacimuthu. *et al.,* 2006).

Traditional botanical preparations represent the primary resource for health care in *Paliyan* society. Tribal herbal practitioners play an important role in helping keep the *Paliyans* healthy. The most common ailments treated are skin problems, burns, wounds, cuts, and diseases of the skin. Other illnesses alleviated by herbal medicines include respiratory infections, coughs, fevers, and colds, and gastrointestinal problems, abdominal pains, stomach aches, throat infections, snake bites, nervous disorders and more (Ignacimuthu *et al.,* 2006).

The following research papers are chosen for this comparative study.

i. Aadhan, K. and Anand, SP. 2018. Ethnomedicinal Plants Utilized by Paliyar's Tribe in Sadhuragiri Hills, Southern Western Ghats, Tamil Nadu, India. *Int. J. Bio. Res.*3(1): 07-15.

ii. Ignacimuthu, S., Ayyanar, M. and SankaraSivaraman, K. 2006. Ethnobotanical Investigations among Tribes in Madurai District of Tamil Nadu (India). *J. Ethnobiol. Ethnomed.* **2**(25): 1-7.

iii. Vikneshwaran, D., Viji, M. and Raja Lakshmi, K. 2008. Ethnomedicinal Plants Survey and Documentation Related to Paliyar Community. *Ethnobot. Leaflets.* **12**: 1108-15.

iv. Shanmugam, S., Gayathri, N., Sakthivel, B., Ramar, S. and Rajendran, K. 2009. Plants Used as Medicine by Paliyar Tribes of Shenbagathope, Virudhunagar District of Tamilnadu, India. *Ethnobot. Leaflets.* **13**: 370-78.

v. Maria Francis Jeffrey Bose, N., Aron, S. and Mehalingam, P. 2014. An Ethnobotanical Study of Medicinal Plants Used by the Paliyans Aboriginal Community in Virudhunagar Distict, Tamil Nadu, India. *Indian J. Tradit Know.*13(3): 613-618.

3. Result and Discussion

Different human ailments faced by the *Paliyans* in their settlements are grouped under 22 ailment categories with their biomedical terms (Table 10.1).

3.1. Comparative Ethno-medico-botany of the *Paliyans* in the Sadhuragiri Hills, Virudhunagar District with the *Paliyans* in and around Madurai District Forest Areas

Comparative Ethno-medico-botany of the *Paliyans* in the Sadhuragiri Hills, Virudhunagar District with the *Paliyans* in and around Madurai District Forest Areas reveals that both use 16 plant species (Table 10.2) to treat different human ailment in 49 and 15 categories respectively. It is observed that out of 16 plant species 9 plant species, namely *Andrographis paniculata* (Others), *Cipodessa baccifera* (Gastro Intestinal Problem), *Dioscorea oppositifolia* (Gastro Intestinal Problem), *Gymnema sylvestre* (Metabolic Disease), *Solanum nigrum* (Gastro Intestinal Problem), *Solanum surattense* (Dental and Oral Problem), *Solanum trilobatum* (Respiratory Problem), *Terminalia chebula* (Respiratory Problem), *Vitex negundo* (Respiratory Problem, Aches and Fever) are used by them to treat ailments in the same category.

Table 10.1: Different Ailments Grouped under different Ailment Categories with their Biomedical Terms

Ailment Categories	*Biomedical Terms*
Aches and Fever	Bodyache, Fever, Influenza, Headache, Malaria
Dental and Oral Problems	Mouth Ulcer, Toothache
Dermatological Problems	Acne, Boils, Eczema, Heel Fissure, Itches, Leprosy, Leucoderma, Ringworm, Scabies, Skin Sores
Gastro Intestinal Problems	Intestinal Worm Infestation, Heart Burn, Colic, Diarrhoea, Dysentery, Nausea, Piles, Constipation, Stomachache, Peptic Ulcer
Gynaecological Problems	Gonorrhea, Leucorrhoea, Menstrual Disorders
Ear and Throat Problems	Earache, Pharyngitis
Haemotological Problems	Haemostatic
Heart Problem	Cardiovascular Disease
Kidney Problem	Hypertension
Liver Problem	Jaundice
Lymphatic Ailment	Buboes
Malignant Disorder	Cancer, Cirrhosis
Metabolic Disease	Diabetes Mellitus
Musculoskeletal Disorder	Arthralgia, Fracture, Gout, Inflammation, Lumbago, Rheumatic Pain (Arthritis), Spasmodic Pain, Swelling
Nerve Ailment	Tetanus
Neurological Problem	Epilepsy, Neurologic Disorders, Paralysis
Others	Burns, Cuts and Wounds, Dog Bite, Hyperthermia, Mammary Hypoplasia, Poisonous Bites (Insect and Scorpion Sting, Snake Bite), Premature Graying of Hair
Otolarynological Problem	Giddiness
Opthalmic Ailment	Common Eye Diseases
Respiratory Problem	Asthma, Bronchitis, Cold, Cough
Sleep Disorder	Insomnia
Urogenital Problem	Endometritis, Urinary Tract Infection, Urolithiasis

It is observed that out of 16 plant species 9 plant species namely *Acorus calamus* (Neurological Problem, Sleep Disorder and Aches and Fever), *Andrographis paniculata* (Others, Kidney Problem, Metabolic Disease, Malignant Disorder, Aches and Fever, Gastro Intestinal Problem and Respiratory Problem), *Anisomeles malabarica* (Aches and Fever, Gastro Intestinal Problem, Nerve Ailment, Musculoskeletal Disorder, Respiratory Problem, Dermatological Problem and Urogenital Problem), *Asparagus racemosus* (Kidney Problem, Malignant Disorder and Others), *Dioscorea oppositifolia* (Gastro Intestinal Problem, Respiratory

Table 10.2: Comparative Ethno-medico-botany of the *Paliyans* in the Sadhuragiri Hills Virudhunagar District with the *Paliyans* in and around Madurai District Forest Areas

Sl. No.	*Botanical Name*	*Part(s) Used*		*Ailments Treated*		*Ailment Categories*	
		Virudhunagar (Sadhuragiri Hills)	*Madurai Forest Areas*	*Virudhunagar (Sadhuragiri Hills)*	*Madurai Forest Areas*	*Virudhunagar (Sadhuragiri Hills)*	*Madurai Forest Areas*
1	*Acorus calamus*	Rhizome	Rhizome	Epilepsy, Insomnia, Bodyache and Hypertension	Pharyngitis	Neurological Problem, Sleep Disorder, Aches and Fever and Kidney Problem	Ear and Throat problems
2	*Alangium salvifolium*	Leaf	Fruit	Heart Burn	Eye Infection	Gastro Intestinal Problem	Ophthalmic Ailment
3	*Andrographis paniculata*	Whole plant	Leaf	Snake Bite, Poisonous Sting, Hypertension, Diabetes, Cancer, Malaria, Influenza, Dysentery and Respiratory Problem	Poison bites, (Scorpion Sting, Snake Bite) and Menstrual Disorders	Others, Kidney Problem, Metabolic Disease, Malignant Disorder, Aches and Fever, Gastro Intestinal Problem and Respiratory Problem	Others and Gynaecological Problem

Sl. No.	Botanical Name	Part(s) Used		Ailments Treated		Ailment Categories	
		Virudhunagar (Sadhuragiri Hills)	Madurai Forest Areas	Virudhunagar (Sadhuragiri Hills)	Madurai Forest Areas	Virudhunagar (Sadhuragiri Hills)	Madurai Forest Areas
4	*Anisomeles malabarica*	Leaf	Stem	Fever, Colic, Boils, Tetanus, Inflammation, Cough, Cold, Stomachache, Itches and Endometritis	Wounds	Aches and Fever, Gastro Intestinal Problem, Nerve Ailment, Musculoskeletal Disorder, Respiratory Problem, Dermatological Problem and Urogenital Problem	Others
5	*Asparagus racemosus*	Tuber	Leaf	Cirrhosis,Mammary Hypoplasia	Heel Fissure	Malignant Disorder,Others	Dermatological Problems
6	*Cipodessa baccifera*	Leaf	Leaf	Dysentery	Diarrhoea	Gastro Intestinal Problem	Gastro Intestinal Problem
7	*Dioscorea oppositifolia*	Root	Rhizome	Diarrhoea, Asthma, Dry Cough and Diabetes Mellitus	Stomachache	Gastro Intestinal Problem, Respiratory Problem and Metabolic Disease	Gastro Intestinal Problem
8	*Gymnema sylvestre*	Leaf	Leaf	Diabetes Mellitus	Diabetes Mellitus and Nervous Disorder	Metabolic Disease	Metabolic Disease and Neurological Problem

Sl. No.	*Botanical Name*	*Part(s) Used*		*Ailments Treated*		*Ailment Categories*	
		Virudhunagar (Sadhuragiri Hills)	*Madurai Forest Areas*	*Virudhunagar (Sadhuragiri Hills)*	*Madurai Forest Areas*	*Virudhunagar (Sadhuragiri Hills)*	*Madurai Forest Areas*
9	*Hemidesmus indicus*	Root	Root and Whole Plant	Mouth Ulcer	Fever	Dental and Oral Problems	Aches and Fever
10	*Piper nigrum*	Seed	Seed	Cold, Cough and Asthma	Pharyngitis	Respiratory Problem	Ear and Throat Problems
11	*Rubia cordifolia*	Leaf and Root	Root	Scorpion Sting andGiddiness	Heel Fissure	Others and Otolstynological Problem	Dermatological Problem
12	*Solanum nigrum*	Fruit	Leaf	Peptic Ulcer	Stomachache	Gastro Intestinal Problem	Gastro Intestinal Problem
13	*Solanum surattense*	Whole Plant	Fruit	Toothache	Toothache	Dental and Oral Problem	Dental and Oral Problem
14	*Solanum trilobatum*	Leaf	Leaf	Cough, Fever, Cold and Asthma	Asthma	Respiratory Problem, Aches and Fever	Respiratory Problem
15	*Terminalia chebula*	Seed	Fruit	Wounds, Leprosy, Inflammation and Cough	Cold, Fever and Stomachache	Others, Dermatological Problem, Musculoskeletal Disorder and Respiratory Problem	Respiratory Problem, Aches and Fever and Gastro Intestinal Problems
16	*Vitex negundo*	Leaf	Leaf	Cold and Headache	Cold, Cough Fever and Headache	Respiratory Problem, Aches and Fever	Respiratory Problem, Aches and Fever

Problem, Metabolic Disease), *Rubia cordifolia* (Others and Otolaryngological Problem), *Solanum trilobatum* (Respiratory Problem, Aches and Fever), *Terminalia chebula* (Others, Dermatological Problem and Respiratory Problem), *Vitex negundo* (Respiratory Problem and Aches and Fever) are used by the *Paliyans* in the Sadhuragiri Hills, Virudhunagar District to treat more than one category of ailment. It is observed that out of 16 plant species, 4 plant species namely *Andrographis paniculata* (Others and Gynaecological Problem), *Gymnema sylvestre* (Metabolic Disease and Neurological Problem), *Terminalia chebula* (Respiratory Problem and Aches and Fever, *Vitex negundo* (Respiratory Problem and Aches and Fever) are used by the *Paliyans* in and around Madurai District Forest Areasto treat more than one category of ailment.

3.2. Comparative Ethno-medico-botany of the *Paliyans* in the Sadhuragiri Hills, VirudhunagarDistrict with the *Paliyans* in the Sirumalai Hills, Dindigul District

Comparative Ethno-medico-botany of the *Paliyans* in the Sadhuragiri Hills, Virudhunagar District with the *Paliyans* in the Sirumalai Hills, Dindigul District reveals that both use 14 plant species (Table 10.3) to treat different human ailments in 40 and 10 categories respectively. It is observed that out of 14 plant species 2 plant species namely *Coleus aromaticus* (Respiratory Problem), *Gymnem asylvestre* (Metabolic Disease) are used by them to treat ailments in the same category. It is observed that out of 16 plant species 4 plant species namely *Catharanthus roseus* (Haemotological Problem and Aches and Fever), *Coleus aromaticus* (Dermatological Problem and Dental and Oral Problem, Musculoskeletal Disorder and Respiratory Problem), *Datura metal* (Gastro Intestinal Problem, Aches and Fever and Musculoskeletal Disorder), *Gloriosa superba* (Other and Dermatological Problem) are used by the *Paliyans* in the Sadhuragiri Hills, Virudhunagar District to treat more than one category of ailment. It is observed that out of 16 plant species 3 plant species namely *Coccinia indica* (Dermatological Problem and Metabolic Disease), *Coleus aromaticus* (Aches and Fever, Respiratory Problem and Gastro Intestinal Problem), *Enicostemma littorale* (Musculoskeletal Disorder and Gynaecological Problem) are used by the *Paliyans* in the Sirumalai Hills, Dindigul District to treat more than one category of ailment.

3.3. Comparative Ethno-medico-botany of the *Paliyans*in the Sadhuragiri Hills, Virudhunagar District with the *Paliyans* in the Shenbagathope, Virudhunagar District

Comparative Ethno-medico-botany of the *Paliyans*in the Sadhuragiri Hills, Virudhunagar District with the *Paliyans* in the Shenbagathope, Virudhunagar District reveals that both use 16 plant species (Table 10.4) to treat different human ailments in 56 and 12 categories respectively. It is observed that out of 16 plant species, namely *Achyranthes aspera* (Dermatological Problem), *Cardiospermum halicacabum* (Musculoskeletal Disorder), *Cissus quadrangualris* (Musculoskeletal

Table 10.3: Comparative Ethno-medico-botany of the *Paliyans* in the Sadhuragiri Hills, Virudhunagar District with the *Paliyans* in the Sirumalai Hills, Dindigul District

Sl. No.	*Botanical Name*	*Part(s) Used*		*Ailments Treated*		*Ailment Categories*	
		Virudhunagar (Sadhuragiri Hills)	*Dindigul (Sirumalai Hills)*	*Virudhunagar (Sadhuragiri Hills)*	*Dindigul (Sirumalai Hills)*	*Virudhunagar (Sadhuragiri Hills)*	*Dindigul (Sirumalai Hills)*
1	*Catharanthus roseus*	Leaf and Root	Root	Haemostatic and Toothache	Itches	Haemotological Problem and Aches and Fever	Dermatological Problem
2	*Cleome viscosa*	Seed	Leaf	Intestinal Worm Infestation	Headache	Gastro Intestinal Problem	Aches and Fever
3	*Coccinia indica*	Leaf	Leaf and Fruit	Peptic Ulcer	Itches, Diabetes Mellitus and Scabies	Gastro Intestinal Problem	Gastro Intestinal Problem
4	*Cocculus hirsutus*	Leaf and Root	Leaf	Stomachache and Rheumatic Pain	Mouth Ulcer	Gastro Intestinal Problem and Musculoskeletal Disorder	Dental and Oral Problem
5	*Coleus aromaticus*	Leaf	Leaf	Skin Sores, Mouth Ulcer, Boils, Swelling and Pharyngitis	Headache, Asthma, Cough, Bronchitis and Stomachache	Dermatological Problem, Dental and Oral Problem, Musculoskeletal Disorder and Throat Problem	Aches and Fever, Respiratory Problem and Gastro Intestinal Problem
6	*Cymbopan citratus*	Whole Plant	Rhizome	Rheumatic Pain	Giddiness	Musculoskeletal Disorder	Otolarynological Problem

Sl. No.	Botanical Name	Part(s) Used		Ailments Treated		Ailment Categories	
		Virudhunagar (Sadhuragiri Hills)	Dindigul (Sirumalai Hills)	Virudhunagar (Sadhuragiri Hills)	Dindigul (Sirumalai Hills)	Virudhunagar (Sadhuragiri Hills)	Dindigul (Sirumalai Hills)
7	*Datura metal*	Fruit	Leaf	Spasmodic Pain, Bodyache and Arthralgia	Boils	Aches and Fever and Musculoskeletal Disorder	Dermatological Problem
8	*Eclipta prostrate*	Whole Plant	Leaf	Jaundice	Premature Graying of Hair	Liver Problem	Others
9	*Enicostemma littorale*	Root	Whole Plant	Toothache	Lumbago and Leucorrhoea	Aches and Fever	Musculoskeletal Disorder and Gynaecological Problem
10	*Euphorbia hirta*	Leaf and Fruit	Leaf and Flower	Leucorrhoea	Mammary Hypoplasia	Gynaecological Problem	Others
11	*Evolvulus alsinoides*	Whole Plant	Leaf	Epilepsy	Fever	Neurological Problem	Aches and Fever
12	*Gloriosa superba*	Rhizome	Tuber and Seed	Poisonous Bite and Itches	Gout	Others and Dermatological Problem	Musculoskeletal Disorder
13	*Gymnema sylvestre*	Leaf	Leaf	Diabetes Mellitus	Diabetes Mellitus	Metabolic Disorder	Metabolic Disorder
14	*Hemidesmus indicus*	Root	Root	Mouth Ulcer	Dysentery	Dental and Oral Problem	Gastro Intestinal Problem

Disorder), *Clitoria ternatea* (Others), *Lantana camara* (Musculoskeletal Disorder), *Leucasas pera* (Dermatological Problem and Respiratory Problem), *Mukkia madraspatna* (Respiratory Problem), *Vernonia cinerea* (Gastro Intestinal Disorder), *Vitex negundo* (Aches and Fever) are used by them to treat ailments in the same category. It is observed that out of 16 plant species 12 plant species namely *Acalypha indica* (Dermatological Problem, Musculoskeletal Disorder and Others), *Aerva lanata* (Gynaecological, Aches and Fever and Kidney Problem), *Albizzia amara* (Gastro Intestinal Problem, Gynaecological Problem, Dermatological Problem and Others), *Cardiospermum halicacabum* (Gastro Intestinal Problem, Lymphatic Ailment, Musculoskeletal Disorder, Neuralogical Problem), *Cissus quadrangualaris* (Others and Musculoskeletal Disorder), *Jatropha curcas* (Musculoskeletal Disorder, Dermatological Problem), *Lantana camara* (Musculoskeletal Disorder, Aches and Fever), *Leucas aspera* (Dermatological Problem and Respiratory Problem), *Mukkiamadraspatna* (Respiratory Problem and Gastro Intestinal Problem), *Tribulusterrestris* (Urogenital Problem and Others), *Vernonia cinerea* (Gastro Intestinal Problem, Aches and Fever, Dermatological Problem and Respiratory Problem), *Vitex negundo* (Respiratory Problem and Aches and Fever) are used by the *Paliyans* in the Sadhuragiri Hills, Virudhunagar District to treat more than one category of ailment. It is observed that out of 16 plant species 11 plant species namely *Acalypha indica* (Respiratory Problem, Ear and Throat Problem and Others), *Achyranthes aspera*(Respiratory Problem and Dermatological Problem and Others), *Cardiospermum halicacabum* (Ear and Throat Problem, Others and Musculoskeletal Disorder), *Cissus quadrangualaris* (Musculoskeletal Disorder and Others), *Hemidesmus indicus* (Aches and Fever, Dermatological Problem and Others), *Jatropha curcas* (Others, Gastro Intestinal Problem and Respiratory Problem), *Lantana camara* (Musculoskeletal Disorder and Aches and Fever), *Leucasaspera* (Dermatological Problem and Respiratory Problem), *Mukkia madraspatna* (Dental and Oral Problem and Respiratory Problem), *Terminalia arjuna* (Gastro Intestinal Problem, Aches and Fever and Dermatological Problem), *Vitex negundo* (Aches and Fever, Gastro Intestinal Problem, Dermatological Problem and Others) are used by the *Paliyans* in the Shenbagathope, Virudhunagar District to treat more than one category of ailment.

3.4. Comparative Ethno-medico-botany of the *Paliyans* in the Sadhuragiri Hills, Virudhunagar District with the *Paliyans* in Some Villages of Virudhunagar District

Comparative Ethno-medico-botany of the *Paliyans* in the Sadhuragiri Hills, Virudhunagar District with the *Paliyans* in some villages of Virudhunagar District reveals (Table 10.5) that both use 14 plants species to treat different human ailments in 39 and 11 categories respectively. It is observed that out of 14 plant species, 7 plant species, namely *Boerhaavia diffusa* (Liver Problem), *Cardiospermum halicacabum* (Musculoskeletal Disorder), *Dodonea viscosa* (Dermatological Problem), *Gymnema sylvestre* (Metabolic Disease), *Mimosa pudica* (Dermatological Problem and Gastrointestinal Problem), *Tridax procumbens* (Others), *Wrightia*

Table 10.4: Comparative Ethno-medico-botany of the *Paliyans* in the Sadhuragiri Hills, Virudhunagar District with the *Paliyans* in Shenbagathope, Virudhunagar District

Sl. No.	*Botanical Name*	*Part(s) Used*		*Ailments Treated*		*Ailment Categories*	
		Virudhunagar (Sadhuragiri Hills)	*Virudhunagar (Shenbagathope)*	*Virudhunagar (Sadhuragiri Hills)*	*Virudhunagar (Shenbagathope)*	*Virudhunagar (Sadhuragiri Hills)*	*Virudhunagar (Shenbagathope)*
1	*Acalypha indica*	Leaf	Whole Plant and Leaf	Ringworm, Rheumatic Pain, Scabies and Wounds	Bronchitis, Earache and Burns	Dermatological Problem, Musculoskeletal Disorder and Others	Respiratory Problem, Ear and Throat Problem and Others
2	*Achyranthes aspera*	Leaf	Whole Plant	Itches	Cough, Skin Eruption, Snake Bite and Dog Bite	Dermatological Problems	Respiratory Problem, Dermatological Problem and Others
3	*Aerva lanata*	Leaf	Leaf and Flower	Gonorrhea, Headache and Urolithiasis	Asthma	Gynaecological Problem, Aches and Fever, Urogenital Problem	Respiratory Problem
4	*Albizzia amara*	Leaf	Leaf and Seed	Diarrhea, Gonorrhea, Itches, Poisonous Bites and Leprosy	Common Eye Diseases	Gastro Intestinal Problem, Gynaecological Problem, Dermatological Problem and Others	Ophthalmic Ailment

Sl. No.	Botanical Name	Part(s) Used		Ailments Treated		Ailment Categories	
		Virudhunagar (Sadhuragiri Hills)	Virudhunagar (Shenbagathope)	Virudhunagar (Sadhuragiri Hills)	Virudhunagar (Shenbagathope)	Virudhunagar (Sadhuragiri Hills)	Virudhunagar (Shenbagathope)
5	*Cardiospermum halicacabum*	Leaf	Whole Plant, Leaf and Root	Nausea, Constipation, Buboes, Common Eye Diseases, Rheumatic Pain and Neurologic Disorders	Earache, Wounds and Rheumatic Pain	Gastro Intestinal Problem, Lymphatic Ailment, Musculoskeletal Disorder and Neurological Problems	Ear and Throat Problem, Others and Musculoskeletal Disorder
6	*Cissus quadrangualaris*	Leaf	Root and Shoot	Wounds and Fracture	Fracture, Burns and Wounds	Others and Musculoskeletal Disorder	Musculoskeletal Disorder and Others
7	*Clitoria ternatea*	Fruit	Leaf and Whole Plant	Insect and Scorpion Bites	Snake Bite	Others	Others
8	*Dodonea viscosa*	Leaf	Leaf	Scabies, Eczema and Acne	Burns and Wounds	Dermatological Problems	Others
9	*Hemidesmus indicus*	Root	Whole Plant	Mouth Ulcer	Fever, Leprosy and Scorpion Sting	Dental and Oral Problems	Aches and Fever, Dermatological Problem and Others
10	*Jatropha curcas*	Seed	Leaf and Seed	Paralysis, Itches	Mammary Hypoplasia, Stomachache and Cough	Neurological Problem and Dermatological Problems	Others, Gastro Intestinal Problem and Respiratory Problem
11	*Lantana camara*	Whole Plant	Whole Plant	Rheumatic Pain and Fever	Malaria and Rheumatic Pain	Musculoskeletal Disorder and Aches and Fever	Musculoskeletal Disorder and Aches and Fever

Sl. No.	Botanical Name	Part(s) Used		Ailments Treated		Ailment Categories	
		Virudhunagar (Sadhuragiri Hills)	Virudhunagar (Shenbagathope)	Virudhunagar (Sadhuragiri Hills)	Virudhunagar (Shenbagathope)	Virudhunagar (Sadhuragiri Hills)	Virudhunagar (Shenbagathope)
13	*Mukkia madraspatna*	Leaf	Root and Fruit	Cough and Stomachache	Toothache and Cough	Respiratory Problem and Gastro Intestinal Problem	Dental and Oral Problem and Respiratory Problem
14	*Tribulus terrestris*	Fruit and Leaf	Fruit	Urinary Tract Infection, Cuts and Wounds	Body ache	Urogenital Problem and Others	Aches and Fever
15	*Veronia cinerea*	Flower and Root	Whole Plant and Root	Intestinal Worm Infestation, Fever, Itches, Cough, Stomachache and Diarrhoea	Piles and Diarrhoea	Gastro Intestinal Problem, Aches and Fever, Dermatological Problem, Respiratory Problem	Gastro Intestinal Problem
16	*Vitex negundo*	Leaf	Leaf, Root and Flower	Cold and Headache	Fever, Peptic Ulcer, Itches, Burn, Dysentery and Headache	Respiratory Problem and Aches and Fever	Aches and Fever, Gastro Intestinal Problem and Dermatological Problem

tinctoria (Dermatological Problem) are used in both place to treat ailments in the same category. It is observed that out of 14 plant species, 5 plant species namely *Anisomeles malabarica* (Aches and Fever, Gastro Intestinal Problem, Dermatological Problem, Musculoskeletal Disorder), *Boerhaavia diffusa* (Respiratory Problem, Dermatological Problem and Liver Problem), *Cardiospermum halicacabum* (Gastro Intestinal Problem, Lymphatic Ailment, Musculoskeletal Disorder), *Mimosa pudica* (Dermatological Problem, Gastro Intestinal Problem and Others), *Pedalium murex* (Respiratory and Urogenital Problem) are used by the *Paliyans* in the Sadhuragiri Hills, Virudhunagar District to treat more than one category of ailment. It is observed that out 14 plant species, 2 plant species namely *Begonia malabarica* (Musculoskeletal Disorder and Kidney Problem), *Mimosa pudica* (Dermatological Problem and Gastro Intestinal Problem) are used by the *Paliyans* in some villages of Virudhunagar District to treat more than one category of ailment.

4. Conclusion

It is evident that there is a well-developed ethno-medicinal system present among the *Paliyans* (indigenous people) in the study Areas where ethno-medicinal surveys were carried out by the investigators. The indigenous people have great expertise with the plants of their own surroundings. The ethno-medicinal information gathered from the comparative ethno-medico botany of the *Paliyans* could be useful for further research in the field of pharmacology. The information could also be useful for the pharmacologists, physicians, phytochemists, botanists and alike interested in the development of alternative therapies and pharmaceuticals. For lesser known plant species, such a secret treasure of ethno-medicinal information could prove beneficial in phyto-pharmacological research for the discovery of new therapeutic drugs. The present study could be of help to develop new drugs based on traditional medicine for effective remedy of different human ailments. The ethno-medicinal information generated from the present study regarding the traditional medicinal plants used by the *Paliyans* need a thorough phytochemical investigation including alkaloid extraction and isolation along with a few clinical trials in future course of time. This could help in creating mass awareness regarding the need for conservation of such valuable medicinal plants and also in the promotion of ethno-medico-botanical knowledge within the region, In view of seeing the exploitation of the biodiversity, traditional knowledge in developing countries like India is eroding at a faster rate. Therefore, it is felt as an urgent need to inventories and record all ethno-medicinal information available in diverse ethnic communities before the traditional practice in health care is completely lost.

5. Recommendations

Based on the results of the present study, the following recommendations are suggested:

- ☆ Resources, especially plant resources are integral to the life of all biota, as they are the primary food producers. Life of world biota is directly or

Table 10.5: Comparative Ethno-medico-botany of the *Paliyans* in the Sadhuragiri Hills, Virudhunagar District with the *Paliyans* in Some Villages of Virudhunagar District

Sl. No.	*Botanical Name*	*Part(s) Used*		*Ailments Treated*		*Ailment Categories*	
		Virudhunagar (Sadhuragiri Hills)	*Virudhunagar (Villages)*	*Virudhunagar (Sadhuragiri Hills)*	*Virudhunagar (Villages)*	*Virudhunagar (Sadhuragiri Hills)*	*Virudhunagar (Villages)*
1	*Abrus precatorius*	Seed	Root	Dysentery	Leucoderma	Gastro Intestinal Problem	Dermatological Problem
2	*Acchyranthes aspera*	Leaf	Leaf	Itches	Dog Bite	Dermatological Problem	Others
3	*Anisomeles malabarica*	Leaf	Leaf	Fever, Colic, Boils, Tetanus, Inflammation, Cough, Cold, Stomachache, Itches and Endometritis,	Wounds	Aches and Fever, Gastro Intestinal Problem, Dermatological Problem, Musculoskeletal Disorder and Urogenital Problem	Others
4	*Begonia malabarica*	Leaf	Aerial Parts	Wounds	Rheumatic Pain and Hypertension	Others	Musculoskeletal Disorder and Kidney Problem
5	*Boerhaavia diffusa*	Whole Plant	Whole Plant	Cough, Itches and Jaundice	Jaundice	Respiratory Problem, Dermatological Problem and Liver Problems	Liver Problem

Sl. No.	Botanical Name	Part(s) Used		Ailments Treated		Ailment Categories	
		Virudhunagar (Sadhuragiri Hills)	*Virudhunagar (Villages)*	*Virudhunagar (Sadhuragiri Hills)*	*Virudhunagar (Villages)*	*Virudhunagar (Sadhuragiri Hills)*	*Virudhunagar (Villages)*
6	*Cardiospermum halicacabum*	Leaf	Whole Plant	Buboes and Rheumatic Pain	Rheumatic Pain	Lymphatic Ailment and Musculoskeletal Disorder	Musculoskeletal Disorder
7	*Coccinia indica*	Leaf	Fruit	Peptic Ulcer	Wounds	Gastro Intestinal Problem	Others
8	*Dodonea viscosa*	Leaf	Leaf	Scabies, Eczema and Acne	Skin disease, Scabies,Eczema and Acne	Dermatological Problem	Dermatological Problem
9	*Gymnema sylvestre*	Leaf	Leaf	Diabetes Mellitus	Diabetes Mellitus	Metabolic Disease	Metabolic Disease
1	*Hemidesmus indicus*	Root	Root	Mouth Ulcer	Hyperthermia	Dental and Oral Problem	Others
11	*Mimosa pudica*	Leaf and Root	Root	Itches, Diarrhoea and Wounds	Scabies and Piles	Dermatological Problem, Gastro Intestinal Problem and Others	Dermatological Problem and Gastro Intestinal Problem
13	*Tridax procumbens*	Leaf	Leaf	Cuts and Wounds	Wounds	Others	Others
14	*Wrightia tinctoria*	Leaf	Leaf	Itches	Scabies	Dermatological Problem	Dermatological Problem

indirectly dependent on plant resources. Thus, the *Paliyans* (indigenous people) in their settlements should be involved in conservation and management plans of plant resources or their indigenous knowledge in their locality.

- ☆ Traditional medicinal plants are central to the indigenous cultures and material needs. Therefore, formal and non-formal education systems should be designed to create positive attitude among the young by integrating into the curricula about the traditional uses of plants in general and medicinal plants in particular.
- ☆ Ethno-medicinal studies should be promoted as they are significant for the discovery of new crude drugs from the reported medicinal plants. Government and particularly NGOs should involve in ethno-medicinal and ethno-pharmaceutical studies to compile and preserve such crucial indigenous knowledge.

6. References

1. Aadhan, K. and Anand, SP. 2018. Ethnomedicinal Plants Utilized by Paliyar's Tribe in Sadhuragiri Hills, Southern Western Ghats, Tamil Nadu, India. *Int. J. Bio Res.* **3**(1): 07-15.
2. Bruce, J. and Meeus, C. 2002. Readers Digest, Sydney.
3. Buenz, E. J., Schnepple, D. J., Bauer, B. A., Elkin, P. L., Riddle, J. M. and Motley, T. J. 2004. Techniques: Bioprospecting Historical Herbal Texts by Hunting for New Leads in Old Tomes. *Trends Pharmacol. Sci.* **25**: 494-498.
4. Cotton, C. M. 1996. Ethnobotany: Principles and Applications. Chichester, UK: John Willey and Sons Ltd. pp. 424.
5. Dahmen, Rev. F. 1908. The Paliyans, a Hill-Tribes of the Palani Hills (South India), Anthropos 3: 19-31.
6. Dutta, B. K. and Dutta, P. K. 2005. Potential of Ethnobotanical in North East India: An Overview. *Indian J. Tradit. Know.* **4**: 7-14.
7. Gardner, P.M. 1972. The Paliyans. In: Bicchieri, M. (ed.) Hunters and Gatherers Today. New York: Holt, Rinehart and Winston. pp. 404-447.
8. Gardner, P. M., 2006. Journey to the Edge in the Footsteps of an Anthropologist. University of Missouri Press. pp. 207.
9. Hiremath, V. T. and Taranath T. C. 2013. Indigenous Traditional Knowledge on Tree Species Survey of Jogimatti Forest, Chitradurga District, Karnataka, India.*J. Ethnobiol. Tradit. Med.* **118**: 222-227.
10. Ignacimuthu, S., Ayyanar, M. and Sankarasivaraman, K. 2006. Ethnobotanical Investigations among Tribes in Madurai District of Tamil Nadu (India). *J. Ethnobiol. Ethnomed.***2**(25): 2-7.

11. Jain, S. K. 1994. Ethnobotany and Research of Medicinal Plants in India. In: Chadwick, D. J. and Marsh, J. (eds.) Ethnobotany and the Search for New Drugs (Ciba Foundation Symposium 185). John Wiley and Sons Ltd., Chichester.
12. Jain, S. K. 1996. Glimpses of Indian Ethnobotany, Oxford and IBH Publishing Co. New Delhi, Bombay, Calcutta.
13. Jain, S. K. and Mudgal, V. 1999. A Handbook of Ethnobotany. Bishen Singh Mahendra Pal Singh Publications, Dehra Dun, India.
14. Jain, D. L., Baheti, A. M., Jain, S. R. and Khandelwal, K. R. 2010. Use of Medicinal Plants among Tribes in Satpuda Region of Dhule and Jalgaon Districts of Maharashtra - An Ethnobotanical Survey. *Indian J. Tradit. Know.* **9**: 152-157.
15. Jansen, P. C. M. 1981. Spices, Condiments and Medicinal plants in Ethiopia, Their Taxonomy and Agricultural Signiûcance. Center for Agricultural Publishing and Documentation, Wageningen, Netherlands. pp. 327.
16. Jeyaprakash, K., Ayyanar, M., Geetha, K. N. and Sekar, T. 2011. Traditional Uses of Medicinal Plants among the Tribal People in Theni District (Western Ghats), Southern India. *Asian Pac. J. Trop. Biomed.* **1**: 20-25.
17. Kala, C. P. 2005. Current Status of Medicinal Plants Used by Traditional Vaidyas in Uttaranchal State of India. *Ethnobot. Res. Appl.* **3**: 267-278.
18. Kibebew, F. 2001. The Status and Availability of Oral and Written Knowledge on Traditional Health Care in Ethiopia. In: (Medhin Zewdu and Abebe Demissieeds). Conservation and Sustainable Use of Medicinal Plants in Ethiopia. Proceeding of the National Workshop on Biodiversity Conservation and Sustainable Use of Medicinal Plants in Ethiopia. *IBCR. Addis Ababa*. pp. 107119.
19. Lee Richard B., Richard Daly and Richard Heywood Daly. 1999. The Cambridge Encyclopedia of Hunters and Gatherers. Cambridge University Press. pp. 61-64.
20. Martin, G. J. 1995. Ethnobotany: A Methods Manual. Chapman and Hall, London. pp. 265-270.
21. Maria Francis Jeffrey Bose, N., Aron, S. and Mehalingam, P. 2014. An Ethnobotanical Study of Medicinal Plants Used by the Paliyans Aboriginal Community in Virudhunagar Distict, Tamil Nadu, India. *Indian J. Tradit. Know.* **13**(3): 613-618.
22. Nithya, N. R. 2014. Globalization and the Plight of Tribals: The Case of Kerala, India. *DJ.* **3**: 727-758.
23. Posey, D. 1999. Ethnobotany: Its Implication and Application. In: Posey, D.A. and Overol, W. L. (Eds.) *Proceedings of the First International Congress of Ethnobiology.* **1**: 1-7.

24. Pushpangadan, P. and Atal, C. K. 1984. Ethno-medico-botanical Investigations in Kerala. Some Primitive Tribes of Western Ghats and Their Herbal Medicines. *J. Ethnopharmacol.* **11**: 59-77.

25. Quanash, N. 1998. Bicultural Diversity and Integrated Health Care in Madagascar. *Nat. Resour.* **30**: 1822.

26. Sajem, A. L. and Gosai, K. 2006. Traditional Use of Medicinal Plants by the Jaintia Tribes in North Cachar Hills District of Assam, Northeast India. *J. Ethnobiol. Ethnomed.* **2**.

27. Schultes, R. E. 1962. The Role of Ethanobotanists in the Search of New Medicinal Plants, Lloydia. 25: 257-266.

28. Shanmugam, S., Gayathri, N., Sakthivel, B., Ramar, S. and Rajendran, K. 2009. Plants Used as Medicine by Paliyar Tribes of Shenbagathope, Virudhunagar District of Tamilnadu, India. *Ethnobot. Leaflets.* 13: 370-78.

29. Solecki, R. and Shanidar, I. V. 1975. A Neanderthal Flower Burial in Northern Iraq. Science. 190: 880881.

30. Thomas Winfred. D. 2003. Evolution of Livelihood Strategies among Geographically Isolated Hill Tribes of South India.

31. Thurston, E. 1909. Castes and Tribes of South India. Government Press. Madras. 4.

32. Vikneshwaran, D., Viji, M. and Raja Lakshmi, K. 2008. Ethnomedicinal Plants Survey and Documentation Related to Paliyar Community. *Ethnobot. Leaflets.* **12**: 1108-15.

33. Yan, X., Murphy, B. T., Hammond, G. B., Vinson, J. A., and Nieto, C. C. 2002. Antioxidant Activities and Antitumor Screening of Extracts from Cranberry Fruit (*Vaccinium macrocarpon*). *J. Agric. Food Chem.* **50**: 5844-5849.

Chapter 11

Healing Traditions: Dermatological Care through Ethnomedicinal Plants of the Irula Tribes in the Nilgiri Hills

Lourdu Pouline S.[1]*, Sathyananth M.*[2]*, Pramely R.*[3] *and Leon Stephan Raj T.*[4]***

[1]*Research Scholar (23211282262008),* [2]*Research Scholar (21111282261014),*
[4]*Assistant Professor and Director, Plant Molecular Biology Research Unit (PMRU),*
Department of Botany, St. Xavier's College (Autonomous), Palayamkottai–627 002
[3]*Assistant Professor,Department of Botany, Kamaraj College, Thoothukudi–628003*
(Affiliated to Manonmaniam Sundaranar University,
Abishekapatti – 627012, Tirunelveli, Tamil Nadu, India
**Corresponding author*

ABSTRACT

Dermatological conditions represent a significant global health challenge, often requiring effective and accessible treatments beyond conventional medicine. Indigenous communities worldwide have developed sophisticated botanical knowledge systems for skin ailments, yet many remain undocumented and at risk of being lost. This study documents the ethnobotanical knowledge of Irula tribes in the Nilgiri Hills of Tamil Nadu regarding medicinal plants used for dermatological conditions. Through semi-structured interviews with traditional healers (vaidyas) across five villages in Kotagiri Taluk, the research identifies 34 plant species employed in treating various skin ailments. The findings reveal sophisticated botanical knowledge including precise preparation methods, application techniques, and strategic multi-plant formulations such as the combination of Acalypha indica, Ruta graveolens, and Curcuma longa for skin rashes. Plants including Azadirachta indica, Aloe vera, and Centella asiatica serve as cornerstone treatments, with their antimicrobial, anti-inflammatory, and wound-healing properties corroborated by contemporary scientific research. This documentation contributes significantly to preserving valuable indigenous knowledge threatened by habitat loss and inadequate recognition of intellectual property rights, while offering potential pathways for developing more accessible and culturally appropriate dermatological interventions.

***Keywords**: Ethnobotany, Dermatological care, Irula tribes, Medicinal plants, Traditional knowledge.*

1. Introduction

The skin is the largest organ of the human body, and serves as the primary interface between internal systems and the external environment. It performs vital functions including protection against pathogens, temperature regulation, sensation, and water retention (Reynolds *et al.,* 2021). This continuous exposure to environmental stressors makes the skin particularly vulnerable to numerous diseases, affecting approximately 900 million people globally and representing the fourth leading cause of nonfatal disease burden worldwide (Goff *et al.,* 2015).

Skin diseases arise from a complex interplay of factors including genetic predisposition, environmental exposures, infectious agents, immunological dysregulation, and lifestyle influences (Barnes *et al.,* 2024). Recent genomic research has identified numerous genetic loci associated with conditions like psoriasis and atopic dermatitis, while environmental elements such as ultraviolet radiation, industrial pollutants, and climate factors significantly contribute to disease initiation and exacerbation (Tsoi *et al.,* 2022; Armstrong and Kricker, 2022). Microbial pathogens–including bacteria, fungi, viruses, and parasites–are responsible for numerous skin infections affecting hundreds of millions globally, while immunological mechanisms underlie many inflammatory and autoimmune skin conditions (Meehan, E. V., and Wang, K. (2022)).

Skin diseases' impact extends beyond physical symptoms to encompass profound psychological, social, economic, and occupational effects. The visible nature of skin conditions contributes to significant psychological distress, with studies demonstrating that the burden of chronic skin diseases can be comparable to that of cancer, arthritis, and heart disease (Balieva, F *et al.,* 2023). Patients frequently experience stigmatization, discrimination, and social isolation, while the economic impact includes substantial direct medical costs, productivity losses, and reduced quality of life (Samela, *et al.,* 2022) Additionally, emerging evidence highlights associations between certain skin diseases and systemic comorbidities including cardiovascular disease and metabolic syndrome (Gelfand *et al.,* 2023).

Against this backdrop, ethnobotanical approaches to skin disease management offer valuable complementary strategies with historical precedence spanning millennia. Archaeological evidence from ancient civilizations documents the use of plant-derived preparations for dermatological conditions, with sophisticated botanical pharmacopeia developed through empirical observation and cultural transmission across generations (Ijinu *et al.,* 2022)

Scientific investigation of traditional plant-based remedies for skin diseases has accelerated in recent decades, with ethnobotanical surveys documenting thousands of plant species used traditionally for dermatological conditions worldwide (Heinrich *et al.,* 2022). Advanced analytical techniques have enabled detailed characterization of bioactive compounds with relevance to dermatological applications, while in vitro, in vivo, and clinical studies have provided mechanistic insights and efficacy data for numerous botanical interventions (Atanasov *et al.,* 2023). Integration of this traditional knowledge with modern pharmaceutical technologies has yielded innovative delivery systems enhancing bioavailability and efficacy, though challenges remain regarding standardization, sustainable sourcing, and appropriate regulatory frameworks (B. Larga *et al.,* 2023).

This chapter examines the intersection of traditional ethnobotanical knowledge and modern dermatological science, exploring the potential of plant-based remedies to address the substantial global burden of skin diseases. By bridging ancient wisdom with contemporary research methodologies, we aim to advance more comprehensive, accessible, and culturally responsive approaches to dermatological care worldwide.

2. Materials and Methods

The ethnobotanical study was conducted in the Nilgiri Hills of Tamil Nadu, known as the 'Queen of Hills.' This region spans diverse ecological zones from evergreen forests to tropical areas, with elevations from 450 meters to 2,636 meters (Chand and Dharmalingam, 2018). Research focused on five small villages in Kotagiri Taluk: Arayur, Kolikari, Mellkoppu, Garikyoor, and Shamakavval. Semi-structured interviews were conducted with Irula vaidyas (traditional healers) who specialize in skin care treatments. Open-ended questionnaires gathered information on medicinal plants, skin ailment remedies, preparation methods, and cultural practices. The study followed ethical standards through prior informed consent and benefit-sharing mechanisms.

Interviews respected Irula protocols to encourage knowledge sharing. Plant specimens were collected with local guides and preserved using standard botanical techniques. The ethnobotanical data was systematically analyzed and cross-referenced with scientific literature to validate pharmacological properties. Despite challenges including language barriers, difficult terrain, and occasional reluctance from knowledge holders, the study successfully documented valuable plant-based skin care knowledge of the Irula tribes, contributing to the preservation of their traditional healing practices.

3. Medicinal Plants of Nilgiris: Traditional Uses and Applications

1) *Acalypha indica* L.

The leaves of Acalypha indica are used in combination with Ruta graveolens leaves and Curcuma longa rhizome. These ingredients are ground together, applied to the body, allowed to dry, and followed by a bath to treat skin rashes. (Suchitra *et al.,* 2023)

2) *Aloe vera* (L.) Burm.f.

The gel is extracted directly from the plant's leaves and applied topically either in its pure form or incorporated into creams and ointments for burns, wounds, dermatitis, and psoriasis, providing moisturizing, anti-inflammatory, and wound-healing benefits (Svitina *et al.,* 2019).

3) *Andrographis paniculata* (Burm.f.) Wall. ex Nees

The leaf paste of Andrographis paniculata is applied externally as a treatment for various skin diseases due to its purported anti-inflammatory and antimicrobial properties. (Sule *et al.,* 2010)

4) *Azadirachta indica* A.Juss.

The leaves and bark (Neem) are typically prepared as pastes, oils, or decoctions and applied topically to treat various skin infections, acne, eczema, and psoriasis, harnessing its antimicrobial, anti-inflammatory, and immunomodulatory properties (Hashmat *et al.,* 2010) Kumari, P., and Singh, M. (2024).

5) *Berberis aristata* DC

Indian Barberry root extract is formulated into ointments and creams and applied to affected areas to address eczema, and psoriasis, and promote wound healing via its antimicrobial, anti-inflammatory, and wound-healing properties (Komal *et al.,* 2011)

6) *Calendula officinalis* L.

Flowers are commonly prepared as infused oils, creams, or tinctures and applied to minor wounds, burns, skin inflammation, and dermatitis to utilize their antimicrobial and tissue regeneration properties (Ejiohuo *et al.,* 2024)

7) *Calotropis gigantea* (L.) W.T.Aiton

Oil and applied to the body as a treatment for skin diseases (Govindasamy *et al.,* 2011)

8) *Catunaregam spinosa* (Thunb.) Tirveng

The root of Catunaregam spinosa is ground into a paste and mixed with lemon juice to create a topical application for leg or toe wounds. This mixture is applied from top to bottom on the affected area. Additionally, this preparation is administered to address lymph node issues (Saini *et al.,* 2019)

9) Centella asiatica (L.) Urb.(Gotu Kola)

The leaves are formulated into extracts, creams, and ointments and applied to wounds, scars, burns, and psoriatic lesions to promote collagen synthesis and improve microcirculation (Sun *et al.,* 2022).(Bylka *et al.,* 2013)

10) Coccinia grandis (L.) Voigt

The leaf paste is applied to skin eruptions once daily until the condition is cured (Hossain *et al.,* 2024)

11) Cullen corylifolium (L.) Medik

Seeds are processed into oils and extracts and applied to affected areas to manage vitiligo, psoriasis, and leukoderma by enhancing melanogenesis and modulating immune responses (Akbar, S., and Akbar, S. (2020).

12) Curcuma longa L

The rhizome of Curcuma longa is ground together with Acalypha indica leaves and Ruta graveolens leaves to create a mixture applied to the body to treat skin rashes. After the mixture dries, a bath is taken to complete the treatment (Aughnet al., 2016)

13) Cynodon dactylon (L.) Pers.

A paste made from one handful of grass and a piece of turmeric (Curcuma longa) is applied to skin diseases.Nagori, B. P., and Solanki, R. (2011).

14) Erythroxylum monogynum Roxb

The stem bark paste is applied to cure skin diseases. This treatment continues for 7 days, with hot water baths taken both morning and evening (Kumar *et al.,* 2019).

15) Flemingia wallichii Wight and Arn.

The leaves are boiled in coconut oil and made into a paste, which is then applied to skin afflictions. During the treatment period of 7 days, it is essential to take a bath in the evening.

16) Hippeastrum puniceum (Lam.) Voss

The bulb of Hippeastrum puniceum is ground into a paste and used in traditional treatments for breast cancer.George, P., and Devi, K. (2023).

17) Ipomoea obscura (L.) Ker Gawl

To treat lumps, a hole is created in the center of an Ipomoea obscura leaf, and coconut or gingelly oil is applied to it before heating. The heated leaf is then placed over the lump, exposing its root. This process reportedly causes the lump to rupture, discharging accumulated matter. For effective treatment, this procedure is repeated for 3 to 4 consecutive days. Ramanathan, S., and Kumar, M. S. (2022).

18) Lawsonia inermis L.:(Henna)

The leaves are ground into a fine powder, mixed with water to form a paste, and applied to affected areas to treat fungal infections, eczema, and wounds by utilizing its antimicrobial, anti-inflammatory, and cooling properties (Gull and Sharma, 2023).

19) Leucas zeylanica (L.) W.T.Aiton

For skin afflictions, the leaf paste is applied in the morning, followed by bathing in the evening. Molla, M. I., and Ela, S. P. (2021)

20) Mimosa pudica L.

For treating skin afflictions, the whole plant is soaked in hot water, and this water is used for bathing purposes. Alternatively, a paste made from the entire plant is applied directly to skin diseases, especially scabies and eczema. Apply once daily for seven days. (Kokane *et al.,* 2009)

21) Moringa oleifera Lam.

The leaves and seeds are prepared as oils, powders, and extracts and applied topically to treat acne, infections, and inflammatory skin conditions through their antimicrobial, anti-inflammatory, and antioxidant effects (Ibrahim and Ahmed, 2023).

22) Nicotiana tabacum L.

Regarding Nicotiana tabacum, it should be noted that applying a mixture of Cassia fistula leaf, tobacco leaf, and human saliva topically to cure skin rashes lacks scientific evidence and may pose potential health risks (Agrawal *et al.,* 2023)

*23) Ocimum tenuifl*orum L.

Holy Basil leaves are crushed into pastes or prepared as infusions and applied directly to affected areas to treat ringworm, eczema, and various skin infections with its antimicrobial and anti-inflammatory properties. One teaspoonful of powdered stem bark from both Tamarindus indica and Terminalia chebula (kadukka) is mixed with coconut oil and applied to eczema-affected areas. This preparation helps in healing and alleviates itching sensations. (Pingale *et al.,* 2012)

24) Phyllanthus emblica L

Fruit extract is processed into serums, creams, and masks and applied to areas with hyperpigmentation, signs of aging, and inflammatory skin conditions, leveraging its high vitamin C content and antioxidant properties (Sharma and Prakash, 2023).

25) *Pongamia pinnata* (L.)

Oil is extracted and applied directly to the skin to manage scabies, ringworm, and other skin infections through its insecticidal, antimicrobial, and anti-inflammatory properties (Kumar and Majumdar, 2022).

26) *Rubia cordifolia* L.

For treating insect bites, including those from beetles, the roots and leaves of Rubia cordifolia are ground together to create a paste. This preparation is then applied directly to the affected area to reduce inflammation and discomfort. It is applied once daily.Patil, R. A., and Jadhav, J. C. (2022).

27) *Ruta graveolens* L.

Ruta graveolens has multiple traditional applications. Its leaves are combined with Acalypha indica leaves and Curcuma longa rhizome, ground together, and applied to the body to treat skin rashes. After the mixture dries, a bath is taken. When grown near houses, the plant is believed to repel snakes. For infants, the leaves are mixed with turmeric and sweet flag for bathing to protect against airborne diseases. Additionally, the leaves are roasted with coconut oil to create a topical paste for treating knee pain and skin diseases. Shahrajabian, M. H. (2024).

28) *Santalum album* L.

(Sandalwood) oil and powder are prepared as pastes by mixing with water or rose water and applied to inflamed or irritated skin to treat acne, inflammatory conditions, and hyperpigmentation with its anti-inflammatory, antimicrobial, and cooling properties (Moy and Levenson, 2022).

29) *Senegalia rugata* (Lam.)

A compound preparation involves grinding the fruits of *Terminalia chebula* and *Senegalia rugata* together with *Grewia tiliifolia* bark. This mixture is applied to the body before bathing to help reduce body heat. Lakshmanan, K., and Narayanan, M. (2022)

30) *Senna septemtrionalis* (Viv.) H.S.Irwin and Barneby

The bark of Senna septemtrionalis is traditionally ground into a paste and applied to the body to detect lumps, particularly in breast tissue. According to local knowledge, if a lump is present, the area where the paste is applied becomes moist or watery, serving as a diagnostic tool. Another preparation involves a mixture of Cassia fistula leaf, tobacco leaf, and human saliva applied topically to treat skin rashes, though it should be noted that this remedy lacks scientific validation and may pose potential risks(Rao *et al.,* 2023)

31) *Sida rhombifolia* L.

The root paste of Sida rhombifolia is mixed with coconut oil and applied to areas affected by scabies. Apply once daily for seven days. A bath should be taken in the evening following each application. (Dinda *et al.,* 2023)

32) *Tamarindus indica* L.

The leaf paste is applied over the entire body to treat eczema. Apply once daily in the morning for seven consecutive days. A bath should be taken in the evening after each application. (Hemshekhar *et al.,* 2011)

33) *Trigonella foenum-graecum* L

Seeds are ground into pastes or steeped in water and applied as compresses to soothe eczema, skin inflammation, wounds, and boils through its anti-inflammatory and antioxidant properties (Ahmadi and Mansourian, 2022)

34) *Withania somnifera* (L.) Dunal:

Root extracts are formulated into creams and ointments and applied to irritated skin to alleviate eczema, psoriasis, and skin inflammation through its anti-inflammatory, antioxidant, and adaptogenic properties (Tripathi and Singh, 2023).

4. Discussion

The ethnobotanical survey of the Nilgiri Hills reveals a rich repository of traditional knowledge regarding plant-based remedies for skin conditions. The documentation of 34 medicinal plant species utilized by Irula vaidyas represents a significant contribution to the preservation of indigenous healing practices while simultaneously offering potential avenues for contemporary dermatological applications.

4.1. Bioactive Compounds and Mechanisms of Action

Analysis of the documented plant species reveals several recurring phytochemical classes that likely contribute to their dermatological efficacy. Many of the identified plants, including *Azadirachta indica, Curcuma longa*, and *Ocimum tenuiflorum*, contain potent antimicrobial compounds such as terpenoids, phenolics, and alkaloids that address the infectious etiology of many skin diseases (Kumari *et al.,* 2024). The anti-inflammatory properties observed in species like *Aloe vera, Calendula officinalis,* and *Withania somnifera* align with contemporary understanding of inflammation as a central pathophysiological mechanism in numerous dermatological conditions (Svitina *et al.,* 2019; Ejiohuo *et al.,* 2024).

Notably, several documented plants exhibit multifaceted dermatological benefits through diverse pharmacological mechanisms. *Centella asiatica*, for instance, promotes wound healing through enhanced collagen synthesis and improved microcirculation (Sun *et al.,* 2020), while *Phyllanthus emblica's* high vitamin C content and antioxidant properties address hyperpigmentation and photoaging (Mehta *et al.,* 2023). This multitargeted approach characteristic of botanical medicines may offer advantages over conventional monotherapeutic pharmaceutical approaches, particularly for complex skin conditions with multiple pathophysiological mechanisms. Traditional plant-based skin treatments from the Irula vaidyas, focusing on their formulation methods and potential modern applications. These remedies include pastes, decoctions, oils, and complex mixtures tailored to specific skin conditions.

Integration with modern dermatology can occur through:

1. Using traditional remedies as leads for drug discovery
2. Building on successful scientific validation (*e.g.*, Aloe vera for wounds)
3. Adopting traditional formulation methods in pharmaceutical development
4. Key limitations include lack of standardized diagnostic criteria, need for toxicological assessment, and insufficient characterization of bioactive compounds.
5. Conservation challenges exist for both the plant species (some facing over harvesting) and the traditional knowledge itself, requiring ethical frameworks and benefit-sharing mechanisms.

Future research should focus on phytochemical characterization of lesser-studied plants, clinical evaluation of traditional formulations, and investigation of traditional preparation techniques using modern pharmaceutical science.

5. Conclusion: Ethnomedicinal Plants of the Irula Tribes for Dermatological Care

This ethnobotanical study in the Nilgiri Hills documents 31 medicinal plant species used by Irula vaidyas for treating skin conditions in the Kotagiri Taluk region. The documented remedies demonstrate sophisticated botanical knowledge, including precise preparation methods and application techniques for various dermatological issues.

Key findings include the Irulas' strategic formulation of multi-plant remedies, such as the combination of *Acalypha indica*, *Ruta graveolens*, and *Curcuma longa* for skin rashes, and their integration of plant materials with vehicles like coconut oil to enhance efficacy. Plants including *Azadirachta indica*, *Aloe vera*, and *Centella asiatica* represent cornerstone treatments with applications corroborated by modern scientific research validating their antimicrobial, anti-inflammatory, and wound-healing properties.

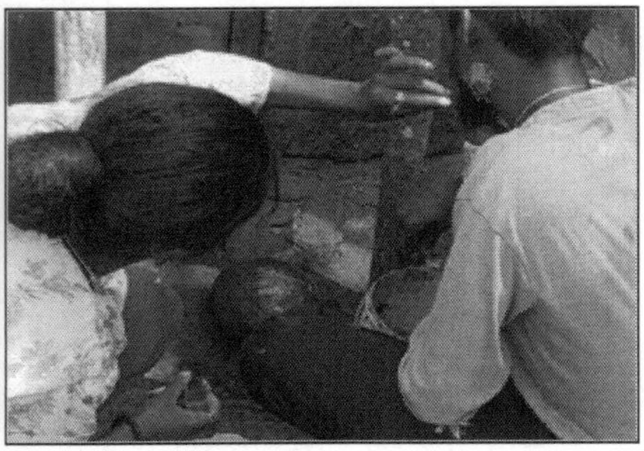

The preservation of this knowledge faces immediate challenges from habitat loss, cultural shifts, and inadequate recognition of indigenous intellectual property. Four critical actions are necessary to safeguard this valuable knowledge system:

1. Systematic documentation of preparation methods and application protocols
2. Scientific validation of traditional formulations through rigorous pharmacological studies
3. Implementation of sustainable cultivation practices for threatened medicinal species
4. Development of equitable benefit-sharing frameworks that recognize Irula intellectual contributions

The ethnopharmacological knowledge of the Irula tribes represents a valuable resource for addressing the global burden of skin diseases. By integrating this traditional knowledge with modern dermatological science, we can develop more effective, accessible, and culturally appropriate interventions while preserving the invaluable botanical heritage of the Nilgiri Hills.

6. References

1. Agrawal, R., Jurel, P., Deshmukh, R., Harwansh, R. K., Garg, A., Kumar, A.,. and Kumarasamy, V. (2024). Emerging trends in the treatment of skin disorders by herbal drugs: traditional and nanotechnological approach. *Pharmaceutics, 16*(7), 869.
2. Ahmadi, R., and Mansourian, M. (2022). Fenugreek (Trigonella foenum-graecum) seed extract in the management of atopic dermatitis: Traditional application and scientific evaluation. Complementary Therapies in Medicine, 70, 102850.

3. Akbar, S., and Akbar, S. (2020). Cullen corylifolium (L.) Medicus (Fabaceae/ Leguminosae) (Syns.: C. corylifolius (L.) Medik; Lotodes corylifolia (L.) Kuntze; Psoralea corylifolia L.; P. patersoniae Schonl.; Trifolium unifolium Forssk). *Handbook of 200 Medicinal Plants: A Comprehensive Review of Their Traditional Medical Uses and Scientific Justifications*, 749-759.

4. Atanasov, A., Grover, L., and Metcalfe, A. Mechanical properties design of pectin-collagen I bioscaffolds for the recapitulation of native skin phenotype traits *in vitro.*

5. B. Larga, J. G., Munabirul, W. T., Moin, A. T., Sarwar Jyoti, M. M., Nasrin, M. S., Al Mueid, M. A.,. and Bonifacio, M. C. (2024). Cutting-edge Bioinformatics strategies for synthesizing Cyclotriazadisulfonamide (CADA) analogs in next-generation HIV therapies. Scientific Reports, 14(1), 29764.

6. Balieva, F., Abebe, D. S., Dalgard, F. J., and Lien, L. (2023). Risk of developing psychiatric disease among adult patients with skin disease: A 9-year national register follow-up study in Norway. Skin health and disease, 3(6), ski2-294.

7. Barnes, B. M., Shyne, A., Gunn, D. A., Griffiths, C. E., and Watson, R. E. (2024). Epigenetics and ultraviolet radiation: Implications for skin ageing and carcinogenesis. Skin Health and Disease, 4(6), e410.

8. Bylka, W., Znajdek–Awi¿eñ, P., Studziñska–Sroka, E., Dañczak–Pazdrowska, A., and Brzeziñska, M. (2014). Centella asiatica in dermatology: an overview. *Phytotherapy research, 28*(8), 1117-1124.

9. Chand, R., and Dharmalingam, M. (2018). The Nilgiris: Ecological and cultural landscapes of a mountain region in South India. Oxford University Press.

10. Dinda, B., Das, N., Dinda, S., Dinda, M., and SilSarma, I. (2015). The genus Sida L.–A traditional medicine: Its ethnopharmacological, phytochemical and pharmacological data for commercial exploitation in herbal drugs industry. *Journal of Ethnopharmacology, 176*, 135-176.

11. Ejiohuo, O., Folami, S., and Maigoro, A. Y. (2024). Calendula in modern medicine: Advancements in wound healing and drug delivery applications. *European Journal of Medicinal Chemistry Reports*, 100199.

12. Ejiohuo, O., Folami, S., and Maigoro, A. Y. (2024). Calendula in modern medicine: Advancements in wound healing and drug delivery applications. European Journal of Medicinal Chemistry Reports, 100199.

13. Gelfand, J. M., Song, W. B., Langan, S. M., and Garshick, M. S. (2024). Cardiodermatology: the heart of the connection between the skin and cardiovascular disease. Nature Reviews Cardiology, 1-18.

14. George, P., and Devi, K. (2023). Cytotoxic and anti-inflammatory properties of Hippeastrum puniceum bulb extracts: Implications for traditional cancer treatments and dermatological applications. Phytotherapy Research, 37(4), 1677-1689.

15. Goff, K. L., Karimkhani, C., Boyers, L. N., Weinstock, M. A., Lott, J. P., Hay, R. J.,. and Dellavalle, R. P. (2015). The global burden of psoriatic skin disease. British Journal of Dermatology, 172(6), 1665-1668.

16. Govindasamy, G. A., Mydin, R. B. S., Sreekantan, S., and Harun, N. H. (2021). Compositions and antimicrobial properties of binary ZnO–CuO nanocomposites encapsulated calcium and carbon from Calotropis gigantea targeted for skin pathogens. *Scientific reports, 11*(1), 99.

17. Gull, A., and Sharma, V. (2023). Lawsonia inermis in dermatology: Beyond cosmetic application to therapeutic potential in fungal infections and inflammatory conditions. Journal of Ethnopharmacology, 310, 116158.

18. Hashmat, I., Azad, H., and Ahmed, A. (2012). Neem (Azadirachta indica A. Juss)-A nature's drugstore: an overview. *Int Res J Biol Sci, 1*(6), 76-9.

19. Heinrich, M., Jiang, H., Scotti, F., Booker, A., Walt, H., Weckerle, C., and Maake, C. (2021). Medicinal plants from the Himalayan region for potential novel antimicrobial and anti-inflammatory skin treatments. Journal of Pharmacy and Pharmacology, 73(7), 956-967.

20. Hemshekhar, M., Kemparaju, K., and Girish, K. S. (2011). Tamarind (Tamarindus indica) seeds: an overview on remedial qualities. *Nuts and seeds in health and disease prevention*, 1107-1114.

21. Hossain, M. S., Jahan, I., Islam, M., Nayeem, J., Anzum, T. S., Afrin, N. A.,. and Hasan, M. K. (2024). Coccinia grandis: Phytochemistry, pharmacology and health benefits. *Clinical Traditional Medicine and Pharmacology, 5*(2), 200150.

22. Ibrahim, H. S., and Ahmed, M. F. (2023**)**. Moringa oleifera in skin disorders: From traditional uses to scientific validation and cosmeceutical applications. Biomedicine and Pharmacotherapy, 159, 114215.

23. Ijinu, T. P., Mohandas, N., Saloni, A., and Pushpangadan, P. (2024). Scientific validation of traditional knowledge: development of herbal drugs and exploration of leads for modern drug discovery. Traditional Medicines in Drug Discovery and Development, 10.

24. Kokane, D. D., More, R. Y., Kale, M. B., Nehete, M. N., Mehendale, P. C., and Gadgoli, C. H. (2009). Evaluation of wound healing activity of root of Mimosa pudica. *Journal of ethnopharmacology, 124*(2), 311-315.

25. Komal, S., Ranjan, B., Neelam, C., Birendra, S., and Kumar, S. N. (2011). Berberis aristata: A review. *Int J Res Ayurveda Pharm, 2*(2), 383-388.

26. Kricker, A., Weber, M. F., Pawlita, M., Sitas, F., Hodgkinson, V. S., Rahman, B.,. and Waterboer, T. (2022). Cutaneous β HPVs, sun exposure, and risk of squamous and basal cell skin cancers in Australia. Cancer Epidemiology, Biomarkers and Prevention, 31(3), 614-624.

27. Kumar, C. D., Anitha, S., and Rao, P. V. D. M. (2019). Evaluation of Potential Phytochemicals and Phyto Pharmacological Activities of Erythroxylum monogynum Roxb. *Biosciences Biotechnology Research Asia, 16*(2), 441.

28. Kumar, T., and Majumdar, A. (2022). Pongamia pinnata in the management of scabies and dermatophytosis: Ethnopharmacological documentation and modern scientific validation. Parasitology Research, 121(3), 1021-1032.
29. Kumari, P., and Singh, M. (2024). The Biochemical Wonders of Azadirachta indica A. Juss: A Comprehensive Review of its Properties and Potential Benefits. *Journal of Stress Physiology and Biochemistry, 20*(2), 90-98.
30. Kumari, R., Sharma, M., Mittal, A., and Goel, G. (2024). Some Versatile Medicinal Plants for Healing Wounds: A Review. *Current Indian Science, 2*(1), E2210299X257271.
31. Lakshmanan, K., and Narayanan, M. (2022). Evaluation of Senegalia rugata fruit extracts for antimicrobial and antifungal activities: Applications in traditional dermatological treatments. Journal of Ethnopharmacology, 298, 115592.
32. Meehan, E. V., and Wang, K. (2022). Interleukin-17 family cytokines in metabolic disorders and cancer. Genes, 13(9), 1643.
33. Mehta, A., Patel, A., Mehta, K., Satvara, K., and Patel, M. (2023). Review of Cosmetic Effects of Natural Bio Functional Ingredient Against Skin Aging.
34. Molla, M. I., and Ela, S. P. (2021). Leucas Zeylanica Is a Bangladeshi Plant with Significant Medicinal Prospect: A Review. *GSC Biol. Pharm. Sci, 16*, 011-018.
35. Moy, R. L., and Levenson, C. (2022). Sandalwood oil in dermatology: Traditional roots with modern branches in acne treatment and skin aging. Journal of Clinical and Aesthetic Dermatology, 15(6), 45-53.
36. Nagori, B. P., and Solanki, R. (2011). Cynodon dactylon (L.) Pers.: A valuable medicinal plant. *Research Journal of Medicinal Plants, 5*(5), 508-14.
37. Patil, R. A., and Jadhav, J. C. (2022). Antimicrobial and anti-inflammatory properties of Rubia cordifolia extracts in insect bite-induced dermatitis: A comprehensive review and in vitro validation. Phytomedicine, 94, 153831.
38. Pingale, S. S., Firke, N. P., and Markandetya, A. G. (2012). Therapeutic activities of Ocimum tenuiflorum accounted in last decade: a review. *J Pharm Res, 5*(4), 2215-2220.
39. Ramanathan, S., and Kumar, M. S. (2022). Anti-inflammatory activity of Ipomoea obscura leaf extracts in the management of subcutaneous cysts: Ethnobotanical perspectives and experimental evidence. International Journal of Dermatology, 61(9), 1122-1135.
40. Rao, P. K., Hasan, S. S., and Bhellum, B. L. (2023). Ethnopharmacological evaluation of Senna septemtrionalis in the treatment of dermatological disorders among indigenous communities of Southern India. Journal of Ethnopharmacology, 301, 115762.
41. Reynolds, G., Vegh, P., Fletcher, J., Poyner, E. F., Stephenson, E., Goh, I.,. and Haniffa, M. (2021). Developmental cell programs are co-opted in inflammatory skin disease. Science, 371(6527), eaba6500.

42. Saini, H., Dwivedi, J., Paliwal, H., Kataria, U., Chauhan, P., and Garg, R. (2019). Anti-inflammatory, analgesic and antipyretic activity of catunaregam spinosa (Thumb.) tirveng extracts. *Journal of Drug Delivery and Therapeutics, 5*(5), 89-94.

43. Samela, T., Sampogna, F., Fabbricatore, M., Fania, L., Papanikolaou, E. S., and Abeni, D. (2022). Skin signs in eating disorders: a literature review. Eating and Weight Disorders-Studies on Anorexia, Bulimia and Obesity, 27(3), 867-879.

44. Shahrajabian, M. H. (2024). A candidate for health promotion, disease prevention and treatment: common rue (Ruta graveolens L.), an important medicinal plant in traditional medicine. *Current reviews in clinical and experimental pharmacology, 19*(1), 2-11.

45. Sharma, P., and Prakash, S. (2023). Phyllanthus emblica extracts in skin aging and hyperpigmentation: Traditional knowledge and modern applications. Journal of Cosmetic Dermatology, 22(4), 1288-1301

46. Suchitra, M. R., Gayathri, R., Parthasarathy, S., and Bhuvaneswari, S. (2023). Evaluation of Phytochemical and in Vitro Anti-inflammatory and Antimicrobial Activity of Acalypha Indica (Kuppaimeni or Harita Manjari) Against Microbes Causing Skin Diseases. *Journal of Pharmaceutical Negative Results, 14*(1), 138-54.

47. Sule, A., Ahmed, Q. U., Samah, O. A., and Omar, M. N. (2010). Screening for antibacterial activity of Andrographis paniculata used in Malaysian folkloric medicine: a possible alternative for the treatment of skin infections. *Ethnobotanical leaflets, 2010*(4), 8.

48. Sun, B., Wu, L., Wu, Y., Zhang, C., Qin, L., Hayashi, M.,. and Liu, T. (2020). Therapeutic potential of Centella asiatica and its triterpenes: a review. *Frontiers in pharmacology, 11*, 568032.

49. Svitina, H., Swanepoel, R., Rossouw, J., Netshimbupfe, H., Gouws, C., and Hamman, J. (2019). Treatment of skin disorders with Aloe materials. *Current pharmaceutical design, 25*(20), 2208-2240.

50. Tripathi, N., and Singh, S. (2023). Withania somnifera in inflammatory skin disorders: Traditional uses, active constituents, and molecular mechanisms. Frontiers in Pharmacology, 14, 1124567.

51. Tsoi, L. C., Patrick, M. T., Shuai, S., Sarkar, M. K., Chi, S., Ruffino, B.,. and Gudjonsson, J. E. (2022). Cytokine responses in nonlesional psoriatic skin as clinical predictor to anti-TNF agents. Journal of Allergy and Clinical Immunology, 149(2), 640-649.

52. Vaughn, A. R., Branum, A., and Sivamani, R. K. (2016). Effects of turmeric (Curcuma longa) on skin health: a systematic review of the clinical evidence. *Phytotherapy Research, 30*(8), 1243-1264.

Chapter 12

Phytochemical Screening and Antioxidant Activity of *Citrus* Fruit Peels

A. Doss[1], *G. Rajeswari*[2] *and N. Ahamed Kabir*[3]

[1]*PG and Research Department of Botany, V. O. Chidambaram College, Thoothukudi, Tamil Nadu*
[2]*Department of Botany, Vellalar College for Women, Erode, Tamil Nadu*
[3]*Research Scholar, PG and Research Department of Botany, V. O. Chidambaram College, Thoothukudi, Tamil Nadu*

1. Introduction

Free radicals play a crucial role in aerobic life and metabolism, acting as essential participants in various biochemical processes (Adesanoye and Farombi, 2014). However, their imbalance or overproduction is implicated in the development of numerous diseases, including cancer (Voituron *et al.,* 2017), Alzheimer's disease, Parkinson's disease (Akinmoladun *et al.,* 2010), inflammatory conditions (Usunobun *et al.,* 2015), lipid peroxidation (Alta *et al.,* 2017), DNA damage (Berma *et al.,* 2013), celiac disease, stroke (Bouzid *et al.,* 2018), cardiovascular diseases (Khan *et al.,* 2013), protein oxidation (Yu *et al.,* 2018), and diabetes (Tufail *et al.,* 2021). Antioxidant activity refers to the capacity of phyto-compounds to protect cellular integrity and function by efficiently eliminating free radicals, preventing lipid peroxidation processes, and halting various types of oxidative harm. Additionally, it serves as a basis for numerous other biological roles, including anti-cancer, anti-inflammatory, and anti-aging effects (Zhuo Zou *et al.,* 2016).

The current growth of functional foods and pharmaceutical products derived from medicinal and edible plants has led to advancements in various aspects of life, including the management of physical disorders, the reduction in the use of synthetic antibiotics, and an increase in life expectancy. Indeed, these plants have been recognized as safe, effective and sustainable sources of natural antioxidants or free radical scavengers, especially in the form of phenolic compounds, like phenolic acids, flavonoids, tannins, stilbenes, and anthocyanins. These phenolic compounds mostly regarded to confer upon the antioxidant properties of medicinal and food plants, playing a significant role in the combating numerous health issues such as cancer, diabetes, aging, cardiovascular, and other degenerative diseases.

Citrus fruits and juices are an important source of bioactive compounds including antioxidants such as ascorbic acid, flavonoids, phenolic compounds and pectins that are important to human nutrition. Flavanones, flavones and flavonols are three types of flavonoids which occur in *Citrus* fruit. The main flavonoids found in citrus species are hesperidine, narirutin, naringin and eriocitrin. An epidemiological study on dietary Citrus flavonoids improved a reduction in risk of coronary heart disease and is attracting more and more attention not only due to their antioxidant properties, but as anti-carcinogenic and anti-inflammatory agents because of their lipid anti-peroxidation effects. The interest in these classes of compounds is due to their pharmacological activity as radical scavengers (Kamran ghasemia *et al.,* 2009).

2. Materials and Methods

2.1. Collection of the Seed Samples

Three different *Citrus* sp. were collected from local markets of Coimbatore, Tamil Nadu dust free samples (Table 12.1, Figure 12.1). Soon after collection, the fruit peels were dried in shade for 20 days and then powdered to get a coarse powder. This powder was stored in air tight container and used for further successive extraction.

2.2. Preparation of Extracts

2.2.1. Aqueous Extraction

25 g of air-dried powder was taken in 100 ml of water in a conical flask, plugged with cotton wool and they were shaken at room temperature for 2 days. After 2 days hours the supernatant was collected and the solvent was evaporated to make the final volume one fourth of the original volume (12) and stored at 4° C in airtight bottles.

2.2.2. Solvent Extraction

25 g of air-dried powder was taken in 100 ml of methanol in a conical flask, plugged with cotton wool and they were shaken at room temperature for 2 days. After 2 days the supernatant was collected and the solvent was evaporated to make the final volume one fourth of the original volume (12) and stored at 4° C in airtight

bottles. In the same conditions was obtained Ethyl acetate extract. The percentage yield of the extracts from each solvent extraction was calculated. The crude extract was crushed into powder and then kept in desiccators.

2.3. Phytochemical Screening

This was carried out according to the methods described by Trease and Evans (1997). Qualification phytochemicals analysis of the crude powder of the three plants for the identification of phytochemicals like as a tannins, alkaloid, steroid, phenols and terpenoid, flavonoid *etc.*

2.3.1. Test for Alkaloids

Five ml of the extract was added to 2ml of HCL. To this acidic medium, 1ml of Dragendroff's reagent was added. An orange or red precipitate produced immediately indicates the presence of alkaloids.

2.3.2. Test for Flavonoids

One ml of the extract, a few drops of dilute sodium hydroxide was added. An intense yellow color was produced in the plant extract, which become colorless on addition of a few drops of dilute acid indicates the presence of flavonoids.

2.3.3. Test for Cardiac Glycosides

The extract was hydrolysed with HCL for few hours on a water bath. To the hydrolysate, 1 ml of pyridine was added and a few drops of sodium nitroprusside solutions were added and then it was made alkaline with sodium hydroxide solution. Appearance of pink to red color shows the presence of glycosides.

2.3.4. Test for Saponins

The extract was diluted with 20ml of distilled water and it was agitated in a graduated cylinder for 15 minutes. The formation of 1 cm layer of foam showed the presence of saponins.

2.3.5. Test for Steroids

One ml of the extracts was dissolved in 10ml of chloroform and equal volume of concentrated sulphuric acid was added by sides of the test tube. The upper layer turns red and sulphuric acid layer showed yellow with green fluorescence. This indicated the presence of steroids.

2.3.6. Test for Tannins

Five ml of the extract and a few drops of 1 per cent lead acetate were added. A yellow precipitate was formed, indicates the presence of tannins.

2.3.7. Test for Triterpenoids

Ten mg of the extract was dissolved in 1ml of chloroform; 1ml of acetic anhydride was added following the addition of 2ml of Conc.H_2SO_4. Formation of reddish violet color indicates the presence of triterpenoids.

2.4. Antioxidant Assay

The antioxidant activity of seed extracts were determined by different *in-vitro* methods such as, the DPPH free radical scavenging assay and reducing power methods.

2.4.1. DPPH Radical Scavenging Activity

DPPH scavenging activity was carried out by the method of Blois, (1958). Different concentrations (1000, 500, 250, 125, 62.5 and 31.2 mg/ml) of dried peel powder of *Citrus sp.* were dissolved in DMSO (dimethyl sulfoxide) and taken in test tubes in triplicates. Then 5 ml of 0.1mM ethanol solution of DPPH (1, 1, Diphenyl-2- Picrylhydrazl) was added to each of the test tubes and were shaken vigorously. They were then allowed to stand at 370 C for 20 minutes. The control was prepared without any extracts. Methanol was used for base line corrections in absorbance (OD) of sample measured at 517nm. A radical scavenging activity was expressed as 1 per cent scavenging activity and was calculated by the following formula. IC 50 Value was also calculated.

$$\text{Radical scavenging activity (per cent)} = \frac{\text{OD Control} - \text{OD Sample x } 100}{\text{OD control}}$$

2.4.2. Reducing Power

Reducing activity was carried out by using the method of *Oyaizu* (1986). Different concentrations (1000, 500, 250, 125, 62.5 and 31.2 mg/ml) of dried peel powder of *Citrus sp.* were dissolved in DMSO (dimethyl sulfoxide) and taken in test tubes in triplicates. To the test tubes 2.5 ml of sodium phosphate buffer and 2.5 ml of 1 per cent Potassium ferric cyanide solution was added. These contents were mixed well and were incubated at 500 C for 20 minutes. After incubation 2.5ml of 10 per cent TCA was added and were kept for centrifugation at 3000rpm for 10 minutes. After centrifugation 5ml of supernatant were taken and to this 5ml of distilled water was added. To this about 1ml of 1 per cent ferric chloride was added and was incubated at 35 0 C for 20 minutes. The O.D (absorbance) was taken at 700nm and the blank was prepared by adding every other solution but without extract and ferric chloride (0.1 per cent) and the control was prepared by adding all other solution but without extract. The reducing power of the extract is linearly proportional to the concentration of the sample

2.5. Quantitative Estimation of Total Phenols and Flavonoids

2.5.1. Total Phenolic Content

Total phenolic contents were determined by Folin Ciocalteu reagent (McDonald *et al.,* 2001). A dilute extract of each plant extract (0.5 ml of 1:10g ml $^{-1}$) or garlic acid (standard phenolic compound) was mixed with Folin Ciocalteu reagent (5ml, 1:10 diluted with distilled water) and aqueous sodium carbonate (4ml, 1 M). The mixtures were allowed to stand for 15 min and the total phenols were determined by colorimetry at 765 nm. The standard curve was prepared using 0, 50, 100, 150, 200, 250 mg/ml solutions of gallic acid in methanol: water (50:50, v/v). Total phenol values are expressed in terms of gallic acid equivalent (mg g-l of dry mass), which is a common reference compound.

2.5.2. Determination of Total Flavonoids

Aluminum chloride colorimetric method was used for flavinoids determination (Chang *et al.,* 2002). Each plant extracts (0.5ml of 1:10 g/ml) in methanol were separately mixed with 1.5 ml of methanol, 0.1ml of 10 per cent aluminum chloride, 0.1ml of 1M potassium acetate and 2.8ml of distilled water. It remained at room temperature for 30 min; the absorbance of the reaction mixture was measured at 415nm with a double beam Perkin Elmer UV/Visible spectrophotometer (USA). The calibration curve was prepared by preparing quercetin solution at concentrations 12.5 to 100g ml^{-1} in methanol.

3. Results

It is well recognized that free radicals are critically involved in various pathological conditions such as cancer, cardiovascular disorders, arthritis, inflammation and liver diseases. Under normal physiological conditions, low concentrations of lipid peroxidation products are found in tissues and cells. In the presence of oxidative stress, more lipid peroxidation products are formed due to cell damage. Cellular antioxidant enzymes such as super oxides dismutase, glutathione peroxidase and catalase normally challenge oxidative stress.

3.1. Collection of Fruits Samples

In the present study, the fruit peels of three different species were collected from the local markets of Coimbatore, South India. After collection, the antioxidant (DPPH and reducing power assay) properties and the levels of poly phenols and flavonoids were analyzed and the results were expressed in the Tables 12.1–12.5.

3.2. Phytochemical Screening

Methanol, Ethyl acetate and aqueous extracts of all the selected fruit tissues were subjected to qualitative organic analyses. Extracts of *Citrus limon, Citrus sinensis* and *Citrus lemettioides* answered positively for terpenoids. Extracts of the plants *Citrus limon* and *Citrus sinensis* answered for alkaloids. Flavonoids were identified in the extracts of *Citrus limon, Citrus sinensis* and *Citrus lemettioides* when tested with Ferric chloride test (Table 12.1).

Table 12.1: Preliminary Phytochemical Screening of Aqueous and Organic Solvent Extracts of *Citrus* Fruit sp.

Phyto-constituents	*Citrus limon*			*Citrus sinensis*			*Citrus limetta*		
	Et.	*EA.*	*Aq.*	*Et.*	*EA.*	*Aq.*	*Et.*	*EA.*	*Aq.*
Tannins	-	-	-	-	-	-	-	-	-
Saponins	-	-	-	-	-	-	-	-	-
Steroids	-	-	-	-	-	-	-	-	-
Triterpenoids	+	+	+	+	+	+	+	+	+
Cardiac glycosides	-	-	-	-	-	-	-	-	-
Alkaloids	-	+	-	-	+	-	-	-	-
Flavonoids	+	+	+	+	+	+	+	+	+

Et.: Ethanol; EA.: Ethyl acetate; Aq.: Aqueous.

3.3. Quantitative Estimation Total Phenols and Flavonoids

Three citrus fruit tissues using water and two different organic solvents were prepared to examine the antioxidant activity and concentrations of Phenols and flavonoids. The extraction solvents of different polarity were used to extract the active substances of different polarity. The concentration of phenols in the examined fruit extracts using the Folin-Ciocalteu reagent was expressed in terms of gallic acid equivalent (Table 12.2). The concentrations of phenols in the examined fruit extracts ranged from 132.6 to 284.92 mg/g. The high concentration of phenols was measured in methanol and aqueous extracts. Ethyl acetate extracts had considerably smaller concentration of phenols. The extracts obtained using more polar solvents had higher concentrations of phenols while the extracts obtained using low polar solvents contained small concentrations (Table 12.2) (Canadanovic *et al.,* 2008).

Table 12.2: Total Phenol Contents in the Fruit Tissue Extracts of *Citrus* sp. Expressed in Terms of mg/100g of Fruit Powder*

Plant Species	*Total Phenol Content*		
	Ethanol	*Ethyl Acetate*	*Water*
Citrus limon	226.29 ± 0.03	176.75 ± 0.12	148.7 ± 0.29
Citrus sinensis	284.92 ± 0.67	182.04 ± 0.24	161.46 ± 0.21
Citrus lemettioides	164.45 ± 0.04	132.6 ± 0.59	149.15 ± 0.24

* Each value in the table was obtained by calculating the average of three analyses ± standard deviation.

The concentration of flavonoids in three different citrus fruit extracts was determined using spectrophotometric method with aluminium chloride. The summary of quantities of flavonoids identified in the tested extracts is shown in

Table 12.3. The concentrations of flavonoids in fruit extracts ranged from 36.14 to 87.17mg/g. High concentrations of flavonoids were measured in ethyl acetate and methanol extracts. The lowest flavonoid concentration was measured in aqueous extracts. The concentration of flavonoids in the extracts depends on the polarity of solvents and the type of plant material used for the extractions. The concentration of flavonoids in plant extracts depends on the polarity of solvents used in the extract preparation (Min and Chun-Zhao, 2005).

Table 12.3: Flavonoids Contents in the Fruit Tissue Extracts of *Citrus* sp. Expressed in Terms of mg/100g of Fruit Powder*

Plant Species	*Flavonoid Content*		
	Ethanol	*Ethyl Acetate*	*Water*
Citrus limon	68.12 ± 0.25	78.03 ± 0.94	46.42 ± 0.43
Citrus sinensis	71.21 ± 0.52	87.17 ± 0.21	59.92 ± 0.11
Citrus lemettioides	64.11 ± 0.05	61.80 ± 0.18	36.14 ± 0.29

* Each value in the table was obtained by calculating the average of three analyses ± standard deviation.

3.4. Antioxidant Activity

3.4.1. DPPH Free Radical Scavenging Activity

The antioxidant activity of the extracts was determined using a DPPH scavenging and reducing power assay. The DPPH assay is often used to evaluate the ability of antioxidants to scavenge free radicals which are known to be a major factor in biological damages caused by oxidative stress. This assay is known to give reliable information concerning the antioxidant ability of the tested compounds.

DPPH is a very stable free radical, unlike *in vitro* generated free radicals, such as hydroxyl radical and superoxide anion, DPPH has the advantage of being unaffected by certain side reactions, such as metal-ion chelation and enzyme inhibition, brought about by various additives. Freshly prepared DPPH solution exhibited a deep purple colour with an absorption maximum at 517 nm. This purple colour generally fades/disappears when antioxidant molecules can quench DPPH free radicals (that is by providing hydrogen atoms or by electron donation, conceivably via a free-radical attack on the DPPH molecule) and converts them into a colourless/bleached product (that is 2, 2-diphenyl-1- hydrazine, or a substituted analogous hydrazine), resulting in a decrease in absorbance at 517 nm band (Suhanya Parthasarathy *et al.*, 2009).

The extracts of all the tested fruit tissues possessed free radical scavenging properties, but to varying degrees, ranging from 12.2 to 92.7 per cent DPPH scavenging. Using the organic solvent extraction, generally Ethyl acetate showed better DPPH scavenging activity. A maximum scavenging activity was offered by Ethyl acetate of *Citrus sinensis* (92.7 per cent), *Citrus limon* (86.4 per cent), *Citrus lemettioides* (72.4) followed by Methanol extract of *Citrus sinensis* (90.78 per cent),

Citrus limon (69.79 per cent), *Citrus lemettioides* (56.1672.4) and water extract of *Citrus sinensis* (79.1 per cent), *Citrus limon* (69.54 per cent) and *Citrus lemettioides* (61.7 per cent) (Table 12.4–12.6).

Table 12.4: DPPH Free Radical Scavenging Assay of Ethyl Acetate extracts of *Citrus* Fruit Tissues*

Plant Species	*Concentration*	*Scavenging Activity (per cent)*	*Standard (Ascorbic Acid)*
Citrus limon	1000	86.4 ± 0.7	71.87 ± 2.21
	500	47.3 ± 1.5	
	250	35.2 ± 3.2	
	125	24.5 ± 2.1	
Citrus sinensis	1000	92.7 ± 2.4	
	500	75.3 ± 0.6	
	250	52.1 ± 0.2	
	125	26.8 ± 2.1	
Citrus lemettioides	1000	72.4 ± 1.6	
	500	54.0 ± 1.8	
	250	36.3 ± 5.9	
	125	21.4 ± 1.4	

* The data are means of triplicate determinations; ± standard errors.

Table 12.5: DPPH Free Radical Scavenging Assay of Methanol Extracts of *Citrus* Fruit Tissues*

Plant Species	*Concentration*	*Scavenging Activity (per cent)*	*Standard (Ascorbic Acid)*
Citrus limon	1000	69.79 ± 4.053	71.87 ± 2.21
	500	60.24 ± 3.719	
	250	54.91 ± 3.666	
	125	38.36 ± 2.573	
Citrus sinensis	1000	90.78 ± 2.536	
	500	80.04 ± 4.774	
	250	72.04 ± 1.910	
	125	61.01 ± 2.474	
Citrus lemettioides	1000	56.16 ± 13.488	
	500	34.68 ± 2.304	
	250	21.21 ± 2.883	
	125	13.03 ± 2.487	

* The data are means of triplicate determinations; ± standard errors.

Table 12.6: DPPH Free Radical Scavenging Assay of Aqueous extracts of *Citrus* Fruit Tissues sp.*

Plant Species	*Concentration*	*Scavenging Activity (per cent)*	*Standard (Ascorbic Acid)*
Citrus limon	1000	69.54 ± 6.07	71.87 ± 2.21
	500	43.3 ± 0.6	
	250	27.39 ± 1.28	
	125	13.5 ± 0.1	
Citrus sinensis	1000	79.1 ± 0.1	
	500	56.4 ± 0.8	
	250	24.59 ± 0.15	
	125	16.1 ± 0.3	
Citrus lemettioides	1000	61.7 ± 5.1	
	500	45.0 ± 1.3	
	250	28.9 ± 0.1	
	125	12.2 ± 1.2	

*The data are means of triplicate determinations; ± standard errors

3.4.2. Reducing Power Assay

The reducing power assay of three different fruit peels of citrus sp. was given in Tables 12.4 and 12.5, which was found to be ranged from 0.179 ± 0.9 - 0.790 ± 2.4 per cent of the dried powder. The reducing power of the fruit tissues increased in a concentration dependent manner. Among the three different sp. the *C. sinensis* (0.790 ± 2.4) have exhibited the highest rate of reduction of ferric ions.

Table 12.7: Reducing Power Assay of Ethyl Acetate Extracts of Fruit Tissues of *Citrus* sp.*

Plant Species	*Concentration*	*Scavenging Activity (per cent)*
Citrus limon	1000	0.686 ± 0.03
	500	0.430 ± 0.01
	250	0.351 ± 0.09
	125	0.263 ± 0.02
Citrus sinensis	1000	0.790 ± 2.4
	500	0.488 ± 1.4
	250	0.395 ± 0.01
	125	0.296 ± 0.01

Plant Species	*Concentration*	*Scavenging Activity (per cent)*
Citrus lemettioides	1000	0.685 ± 0.002
	500	0.455 ± 0.03
	250	0.297 ± 0.01
	125	0.179 ± 0.9

*The data are means of triplicate determinations; ± standard errors.

Table 12.8: Reducing Power Assay of Methanol Extracts of Fruit Tissues of *Citrus* sp.*

Plant Species	*Concentration*	*Scavenging Activity (per cent)*
Citrus limon	1000	0.610 ± 2.1
	500	0.430 ± 0.04
	250	0.333 ± 0.001
	125	0.227 ± 1.2
Citrus sinensis	1000	0.722 ± 1.7
	500	0.523 ± 0.01
	250	0.318 ± 0.03
	125	0.198 ± 0.3
Citrus lemettioides	1000	0.549 ± 0.04
	500	0.408 ± 0.01
	250	0.336 ± 0.01
	125	0.221 ± 0.8

* The data are means of triplicate determinations; ± standard errors.

Table 12.9: Reducing Power Assay of Aqueous Extracts of Fruit Tissues of *Citrus* sp.*

Plant Species	*Concentration*	*Scavenging Activity (per cent)*
Citrus limon	1000	0.481
	500	0.243
	250	0.154
	125	0.095
Citrus sinensis	1000	0.593
	500	0.282
	250	0.124
	125	0.094

Plant Species	*Concentration*	*Scavenging Activity (per cent)*
Citrus lemettioides	1000	0.313
	500	0.216
	250	0.157
	125	0.084

*The data are means of triplicate determinations; ± standard errors

4. Discussion

Free radicals contribute to more than one hundred disorders in humans including atherosclerosis, arthritis, ischemia and repercussion injury of many tissues, central nervous system injury, gastritis, cancer and AIDS (Pourmorad *et al.,* 2006). Antioxidants through their scavenging power are useful for the management of those diseases. DPPH stable free radical method is an easy, rapid and sensitive way to survey the antioxidant activity of a specific compound or plant extracts. Antioxidants, both exogenous and endogenous, whether synthetic or natural, can be effective in prevention of the free radical formation by scavenging or by effecting promotion of their decomposition and suppression of such disorders (Souri *et al.,* 2008). There is growing interest towards natural antioxidants from herbal sources.

Plant materials containing phenolic constituents are increasingly of interest for probing as they retard oxidative degradation of lipids and thereby improve the quality and nutritional value of food. The importance of the antioxidant constituents of plant material in the maintenance of health and protection from coronary heart disease and cancer has been well recognized (Khalil *et al.,* 2007). It has been recognized that flavonoids show antioxidant activity and their effects on human nutrition and health are considerable. The mechanisms of action of flavonoids are through scavenging or chelating process (Kessler *et al.,* 2003). Phenolic compounds are a class of antioxidant agents which act as free radical terminators (Shahidi and Wanasundara, 1992).

4.1. Phytochemical Screening

All plant parts synthesize some chemicals by themselves, to perform their physiological activities. In our present study, the investigated plants have exhibited different kinds of secondary metabolites. The medicinal value of these secondary metabolites is due to the presence of chemical substances that produce a definite physiological action on the human body. The most important of these substances include, alkaloids, glucosides, steroids, flavonoids, fatty oils, resins, mucilages, tannins, gums, phosphorus and calcium for cell growth, replacement, and body building (Kubmarawa *et al.,* 2008).

4.2. Antioxidant Activity

4.2.1. DPPH Free Radical scavenging activity

The α, α,diphenyl-β-picrylhydrazyl (DPPH) a stable nitrogen centered free radical, has been used to evaluate the antioxidant activity of natural products by measuring the radical quenching capacity in a relatively short period of time. All the presently studied crude extracts were found to exhibit more effective free radical inhibition activity against DPPH.

The free radical inhibition activity of crude methanolic extract of citrus fruit peel was ranged between 12.22 ± 1.2 and 92.7 ± 2.4 per cent, which is in agreement with that of the previous reports on Banana (*Musa*, AAA cv. Cavendish (Matook Saif Mokbel and Fumio Hashinaga, 2005); *Hylocereus undatus* (Nurliyana *et al.,* 2010); Apple (Maria Leontowicz *et al.,* 2003). Among the three different samples, the *C. sinensis* showed the highest free radical scavenging activity.

4.2.2. Reducing Power Assay

Reducing power assay is often to evaluate the ability of natural antioxidant to donate electron or hydrogen. Samples with high reducing power were reported to have a better ability to electrons. It has been widely accepted that the higher level of absorbance at 700nm indicates greater reducing power of the test samples (Duh *et al.,* 1997).

Four different concentrations (1000, 500, 250 and 125μg/ml) of the Peel tissue extracts of citrus fruits were found to exhibit higher level of reducing power in a dose-dependent manner (Tables 12.4 and 12.5). This is in agreement with an earlier report on *Hylocereus undatus* (Nurliyana *et al.,* 2010).

4.3. Polyphenols and Flavonoids

Phenolic and other phytochemical antioxidants found in fruits, vegetables and legumes are bioactive compounds capable of neutralizing free radicals and may play a role in the prevention of certain diseases. Also, dietary supplements and food fortification may be an alternative route to the consumption of minor plant components that may have health effects. The high phenol and flavonoids contents of Peels and tissue of *C.sinesis* may cause high antioxidant activity of this plant. Phenols and polyphenolic compounds, such as flavonoids, are widely found in food products derived from plant sources, and they have been shown to possess significant antioxidant activities The correlation between total phenol contents and antioxidant activity has been widely studied in different foodstuffs such as fruit and vegetables (Jayaprakasha *et al.,* 2008). As reported, antioxidant activity of fruits and vegetables significantly increases with the presence of high concentration of total polyphenol content.

5. Conclusion

Perspectives of utilization of fruits and vegetables wastes-originated antioxidants are relatively big, but up to now little recognized. The research accomplished indicated the fruit peels as potential source of natural substances with antioxidant properties. It was demonstrated that among numerous already identified polyphenols compounds, tannins constitute an essential fraction in the peels of citrus fruits. In many cases, the fruit seeds and peels are the waste products of technological processes, hence their re-using as the antioxidant source, could bring measurable economical profits and contribute to reduction of pollutions introduced by fruit and vegetable industries into the environment. It is necessary to consider both environmental (waste management, protection against pollution) and economical aspects (extraction profitability).

We have collected three different types of citrus sp. from local markets of Coimbatore District, South India. The aforesaid species were analyzed for their phytoconstituents and antioxidant activity. The results of the present study indicated that the citrus species have good antioxidant activity with high level flavonoids comparable with that of other fruits. Among the three different species, *Citrus sinensis* was found to possess a rich source of poly phenols and flavonoids. Concluding, it should be stressed that the peels and seeds of many fruits can be considered as a potential source of different antioxidant components, which are not exploited at the moment, but could find practical application in many industrial branches.

6. References

1. Adesanoye, O.; Farombi, E. In Vitro Antioxidant Properties of Methanolic Leaf Extract of Vernonia Amygdalina Del. Nigerian J. Physiological Sci. 2014, 29, 93–101.
2. Akinmoladun, A. C.; Obuotor, E. M.; Farombi, E. O. Evaluation of Antioxidant and Free Radical Scavenging Capacities of Some Nigerian Indigenous Medicinal Plants. J. Med. Food. 2010, 13, 444–451. DOI: 10.1089/jmf.2008.0292.
3. Atta, A.; Mustafa, G.; Sheikh, M. A.; Shahid, M.; Xiao, H. The Biochemical Significances of the Proximate, Mineral and Phytochemical Composition of Selected Vegetables from Pakistan./Mat. Sc. Pharm. 2017, 1(1), 06–09.
4. Berma, S.; Singh, D.; Verma, P.; Soni, D.; Gupta, A.; Nema, R. Review Article Boerhavia Diffusa: An Important Medicinal Plant and Their Phytochemistry. CMBT J. Sci.Tech. 2013, 1(1), 01–04.
5. Blois, MS., 1958. Antioxidant determinations by the use of stable free radical. *Nature* **81**: 1199-2000. (Original not seen).
6. Bouzid, M. A.; Filaire, E.; Matran, R.; Robin, S.; and Fabre, C., *et al.,* Lifelong Voluntary Exercise Modulates age-related Changes in Oxidative Stress. Int. J. Sports. Med. 2018, 39(1), 21–28. https: //doi.org/10.1055/s-0043-119882
7. Canadanovic, BJ., Cetkovic, G., Dilas, S., Tumbas, V., Bogdanovic, G., Mandic, A., Markov, S., Cvetkovic, D. and Canadanovic, V., 2008. Radical scavenging, antibacterial, and antiproliferative activites of *Melissa officinalis* L. extracts. *J. Med. Food.,* **11**: 133-143.

8. Chang, C., Yang, M., Wen, H. and Chern, J. 2002. Estimation of total flavonoid content in propolis by two complementary colorimetric methods. *J. Food Drug Anal.* **10**: 178-182.

9. Duh,PD.,Yen,WJ.,Du,PC. and Yen,G.C., 1997. Antioxidant activity of mung bean hulls. *J.Am.Oil.Chem.Soc.***74**: 1059-1063.

10. Jayaprakasha, GK., Girennavar, B. and Patil, BS., 2008. Radical scavenging activities of Rio Red grapefruits and Sour orange fruit extracts in different *in vitro* model systems. *Bioresour. Technol.* **99(10):** 4484-4494.

11. Kamran ghasemia, yosef ghasemia and mohammad ali ebrahimzadeh, 2009. Antioxidant activity, phenol and flavonoid contents of 13 citrus species peels and tissues. *Pak. J. Pharm. Sci.* **22(3)**: 277-281

12. Kessler, M., Ubeaud, G. and Jung, L., 2003. Anti and pro-oxidant activity of rutin and quercetin derivatives. *J. Pharm. Pharmacol.* **55**: 131-142.

13. Khalil, MY., Moustafa, AA. and Naguib, NY., 2007.Growth,phenolic compounds and antioxidant activity of some medicinal plants grown under organic farming condition. *World J. Agri. Sci.*, **3(4):** 451-457.

14. Khan, M. S.; Ansari, I. A.; Ahmad, S.; Akhtar, F.; Hashim, A.; Srivastava, A. K. Chemotherapeutic Potential of Boerhaavia Diffusa Linn; A Review. J. Applied Pharmaceutical Sci. 2013, 3(1), 133–139.

15. Kubmarawa, D., Khan, ME., Punah, AM. and Hassan, 2008. Phytochemical screening and antibacterial activity of extracts from *Parkia clappertoniana* keay against human pathogenic bacteria. *J. Med. Pl. Res.* **2(12)**: 352-355.

16. Maria Leontowicz, Shela Gorinstein, Hanna Leontowicz, Ryszard Krzeminski, Antonin Lojek, Elena Katrich, Milan Cyiazy, Olga Martin-Belloso, Robert Soliva-Fortuny, Ratiporn Haruenkit, and Simon Trakhtenberg, 2003. Apple and Pear Peel and Pulp and Their Influence on Plasma Lipids and Antioxidant Potentials in Rats Fed Cholesterol-Containing Diets. *J. Agric. Food Chem.***51**: 5780-5785

17. Matook Saif Mokbel and Fumio Hashinaga, 2005. Antibacterial and Antioxidant Activities of Banana (*Musa*, AAA cv. Cavendish) Fruits Peel. *Am.J. Biochem. Biotechnol.***1 (3)**: 125-131.

18. McDonald, S., Prenzler, P.D., Autolovich, M. and Robards, K., 2001. Phenolic content and antioxidant activity of olive extracts. *Food Chem.* **73**: 73-84.

19. Min, G. and Chun-Zhao, L., 2005. Comparison of techniques for the extraction of flavonoids from cultured cells of *Saussurea medusa* Maxim. *World J. Microb. Biot.*, **21**: 1461-1463.

20. Nurliyana, R., Syed Zahir, I., Mustapha Suleiman, K., Aisyah, MR. and Kamarul Rahim, K., 2010. Antioxidant study of pulps and peels of dragon fruits: a comparative study. *Int. Food Res. J.* **17**: 367-375

21. Oyaizu, M., 1986. Studies on products of browning reaction: Antioxidant activities of products of browning reaction prepared from glucosamine. *Jap. J. Nut.* **44**: 307-15.

22. Pourmorad, F., Hosseinimehr, SJ. and Shahabimajd, N., 2006. Antioxidant activity, phenol and flavonoid contents of some selected Iranian medicinal plants. *Afr. J.Biotech.* 5 (11): 1142-1145.

23. Shahidi, F. and Wanasundara, PKJPD., 1992. Phenolic antioxidants. *Crit. Rev. Food Sci. Nut.* **32**: 67-103.

24. Souri, E., Amin, G., Farsam, H. and Barazandeh Tehrani, M., 2008. Screening of antioxidant activity and phenolic content of 24 medicinal plant extracts. *DARU.* **16**: 283-287.

25. Suhanya Parthasarathy, Juzaili Bin Azizi, Surash Ramanathan, Sabariah Ismail, Sreenivasan Sasidharan, Mohd Ikram Mohd. Said and Sharif Mahsufi Mansor, 2009. Evaluation of Antioxidant and Antibacterial Activities of Aqueous, Methanolic and Alkaloid Extracts from *Mitragyna Speciosa* (Rubiaceae Family) Leaves. *Molecules,* **14**: 3964-3974; doi: 10.3390/molecules14103964

26. Trease, G. E. and Evans, W. C. 1997. *Pharmacognosy,* Ed. XIV, Mac Millian Publisher Ltd., London. pp. 568 – 578.

27. Tufail, T.; Ijaz, A.; Noreen, S.; Arshad, M. U.; Gilani, S. A.; Bashir, S.; Din, A.; Shahid, M. Z.; Khan, A. A.; Khalil, A. A., *et al.,* Pathophysiology of Obesity and Diabetes. In Dietary Phytochemicals; Egbuna, C., Hassan, S., Eds.; Springer: Cham, 2021; pp 29–42. DOI: 10.1007/978-3-030-72999-8_2.

28. Usunobun, U.; Okolie, N. P.; Anyanwu, O. G.; Adegbegi, A. J. J.; Egharevba, M. E. Phytochemical Screening and Proximate Composition of Annona Muricata Leaves. Eur J.Botany, Plant Sci and Phytology. 2015, 2, 18–28.

29. Voituron, Y.; Josserand, R.; Le Galliard, J.-F.; Haussy, C.; Roussel, D.; Romestaing, C.; Meylan, S., *et al.,* Chronic Stress, Energy Transduction, and free-radical Production in a Reptile. Oecologia. 2017, 185(2), 195–203. https: //doi. org/10.1007/s00442-017-3933-1

30. Yu, Y.; Gao, Q.; Xia, W.; Zhang, L.; Hu, Z.; Wu, X.; Jia, X., *et al.,* Association between Physical Exercise and Biomarkers of Oxidative Stress among middle-aged and Elderly Community Residents with Essential Hypertension in China. Biomed Res. Int. 2018, 2018, 4135104. DOI: 10.1155/2018/4135104.

31. Zhuo Zou, Wanpeng Xi, Yan Hu, Chao Nie, Zhiqin Zhou, 2016. Antioxidant activity of Citrus fruits. Food Chemistry.196: 885-896

Chapter 13

Phytochemical Screening of Traditional Medicinal Plants in Tuticorin

N. Ahamed Kabir[1], G. Rajeswari[2] and A. Doss[3]

[1]Research Scholar, PG and Research Department of Botany, V.O. Chidambaram College, Thoothukudi, Tamil Nadu
[2]Department of Botany, Vellalar College for Women, Erode, Tamil Nadu
[3]PG and Research Department of Botany, V.O. Chidambaram College, Thoothukudi, Tamil Nadu

1. Introduction

Antibiotics are substances that either inhibit the growth of bacteria (bacteriostatic agents) or eliminate them (bactericidal agents). These substances effectively lower or eradicate bacterial populations by obstructing essential bacterial cellular functions. However, the presence of antimicrobial resistance has reached critical levels globally and is a significant concern due to the adaptation mechanisms exhibited by microorganisms. Since ancient times, humans have searched for new drugs in nature, particularly among plants, leading to the use of numerous medicinal plants with curative properties to treat various diseases. The World Health Organization (WHO) states that traditional plant-based medicines are the primary source of healthcare for over 80 per cent of the population in developing and underprivileged nations (Rao *et al.,* 2016). Moreover, the WHO has made an effort to identify all internationally used medicinal plants and recognized over 20,000 species (Kareli *et al.,* 2022). In India, it is reported that nearly 95 per cent of prescriptions in traditional systems such as f Unani, Ayurveda, Homeopathy and Siddha are plant-based products.(Preeti Kumar *et al.,* 2017). Medicinal plants are a valuable resource, as they offer a wide range of potential therapeutic compounds that are both diversified and reasonably safe, compared to manufactured pharmaceuticals (Rao *et al.,* 2016). These bioactive compounds of plants include alkaloids, saponins, steroids, terpenes, glycosides, flavonoids, tannins and phenolic compounds. There is a rapidly increasing interest among scientists in identifying secondary metabolites generated by plants and assessing their potential for disease treatment (Marquardt *et al.,* 2020; Mapfuman *et al.,* 2022). This is due to the fact that novel compounds derived from medicinal plants can play a crucial role in the creation and production of both synthesized and natural molecules for contemporary therapeutic systems (Shibula *et al.,* 2015).

Vitex negundo L. is a shrub used in traditional medicine for various diseases due to its hepatoprotective, antioxidant, anti-inflammatory, analgesic, and antifungal properties. It contains various chemical compounds, including polyphenolic, terpenoids, glycosidic iridoids, and alkaloids. The plant has been isolated from various sources, including Vitegnoside, negundoside, agunside, and herbacetin 3-rhamnoside (Meena *et al.,* 2010).

Ocimum tenuiflorum L., often known as Tulasi, is a perennial plant in the Lamiaceae family that is native to the Indian subcontinent but is widely grown across Southeast Asia's tropics. It is a fragrant plant from the basil family. Tulasi is an antioxidant that helps to avoid the indications of aging. Basil or Tulasi protects against oxidative damage, lowers free radicals, and regulates antioxidant enzymes. It is beneficial against a wide range of disorders, including cancer, bronchitis, helmithiasis, ulcers, inflammation, and rheumatism (Cohen, 2014; Dharmani and Palit, 2006; Pandey *et al.,* 2014; Singh and Chaudhuri, 2018; Singh and Majumdar, 1999). Numerous preclinical investigations have assessed tulsi's anti-ulcerogenic properties, which are thought to be caused by the presence of fixed oil eugenol (Dharmani *et al.,* 2004). In a variety of ulcer models in rats and guinea pigs, it has been discovered that fixed oil derived from *O. tenuiflorum* reduces ulcer index as well as total and free acidity. Singh and Majumdar (1999) identified histamine antagonistic effects, LOX inhibition, and anti-secretory effects as potential pathways.

Ocimum gratissimum L., also referred to as smell leaf, is one of the newly found medicinal plants that may be used as a source for a new medication or as an alternative therapy for a number of illnesses. It is a fragrant perennial herbaceous plant that is widely grown and economically successful. It is found in Africa, Asia, and South America and is a member of the Lamiaceae family (Tanko *et al.,* 2008; Akara *et al.,* 2021). It may be added to fish, meat, soups, and stews as a natural flavoring, seasoning, or vegetable. A number of illnesses, including cough, pneumonia, fever, inflammation, anemia, diarrhea, aches, and bacterial and fungal infections, are also treated with it in traditional medicine (Akara *et al.,* 2021).

Azadirachta indica and *Melia azedarach* are two closely related species of Meliaceae, known for their medicinal properties. *Azadirachta indica*, also known as Indian Neem or Persian lilac, is a small-to-medium-sized deciduous tree with a spreading, dense, and dark green crown. Its bark is dark brown and smooth, and its leaves are alternate, with short stalked, thin, hairless, dark green leaflets. The flowers are white with purple stripes and have a typical fragrance. The fruit or berries are yellow, round, smooth, and fleshy, with dried fruits having 4 to 5 seeds. *M. azedarach* Linn. is native to tropical Asia and is widespread and naturalized in most tropics and subtropical countries. The leaves have insecticidal properties, while the seed oil is the most active medicinal product of the plant. It is used as an antiseptic for sores, ulcers, rheumatism, skin diseases, malaria fever, and leprosy (Ramya *et al.*, 2009; Zakir-ur-Rahman *et al.*, 1991).

Tamarindus indica L. (Tamarind), a tropical fruit found in Africa and Asia, is highly valued for its pulp, which has a sweet acidic taste due to high tartaric acid and reducing sugars. It is used for seasoning, flavored confections, curries, sauces, and juices, and is a major ingredient in beverages. Tamarind seeds are becoming an alternative source of proteins and are rich in essential minerals. Seed pectin can form gels over a wide pH range. Tamarind leaves and flowers are used in various dishes, including curries, salads, stews, and soups. They are rich in vitamin C and α-carotene, and have anti-oxidant, anti-inflammatory, anti-microbial, and anti-fungal activity. Tamarind is also extensively used in traditional medicine, with its traditional uses, phytochemistry, and pharmacognosy reviewed to provide a particular orientation to its value in sub-Sahara Africa (De Caluwé *et al.*, 2010).

2. Materials and Methods

2.1. Plant Materials

Fresh plant parts were collected randomly from the gardens and villages of Tuticorin district, Tamil Nadu from the natural stands. The botanical identity of these plants was confirmed by Botanical Survey of India (Southern Circle), Coimbatore, Tamil Nadu (Plate 13.1).

***Vitex negundo* L.**

***Ocimum tenuiflorum* L.**

Ocimum gratissimum **L.**

Melia azadirachta **L.**

Tamarindus indica **L.**

Plate 13.1: Medicinal Plants.

2.2. Preparation of Extracts

2.2.1. Aqueous Extraction

100 grams of dried powder were extracted in distilled water for 6 h at slow heat. Every 2 h it was filtered through Whatman no.1filter paper and centrifuged at 5000 g for 15 min. The supernatant was collected. This procedure was repeated twice and after 6 h the supernatant was concentrated to make the final volume one-fifth of the original volume.

2.2.2. Solvent Extraction

100 grams of dried plant powdered samples were extracted with 200 ml of methanol kept on a rotary shaker for 24 h. Thereafter, it was filtered and centrifuged at 5000 g for 15 min. The supernatant was collected and the solvent was evaporated to make the final volume one-fifth of the original volume. It was stored at 4°C in airtight bottles for further studies, *viz.* antimicrobial, antioxidant, anticancer and phytochemical analysis (Figure 13.1).

Figure 13.1: Crude Extracts (Ethanol and Aqueous).

2.3. Bioactive Compounds Screening

Phytochemical screening of crude extracts of all the plants were carried out according to the methods described by Trease and Evans (1997). Qualification phytochemicals analysis of the crude powder of the samples for the identification of phytochemicals like as a tannin, alkaloid, steroid, phenols and terpenoid, flavonoid *etc.*

2.3.1. Test for Alkaloids

Five ml of the extract was added to 2 ml of HCL. To this acidic medium, 1 ml of Dragendroff's reagent was added. An orange or red precipitate produced immediately indicates the presence of alkaloids.

2.3.2. Test for Flavonoids

One ml of the extract, a few drops of dilute sodium hydroxide was added. An intense yellow color was produced in the plant extract, which become colorless on addition of a few drops of dilute acid indicates the presence of flavonoids.

2.3.3. Test for Saponins

The extract was diluted with 20 ml of distilled water and it was agitated in a graduated cylinder for 15 minutes. The formation of 1 cm layer of foam showed the presence of saponins.

2.3.4. Test for Steroids

One ml of the extracts was dissolved in 10 ml of chloroform and equal volume of concentrated sulphuric acid was added by sides of the test tube. The upper layer turns red and sulphuric acid layer showed yellow with green fluorescence. This indicated the presence of steroids.

2.3.5. Test for Cardiac Glycosides

The extract was hydrolysed with HCL for few hours on a water bath. To the hydrolysate, 1 ml of pyridine was added and a few drops of sodium nitroprusside solutions were added and then it was made alkaline with sodium hydroxide solution. Appearance of pink to red color shows the presence of glycosides.

2.3.6. Test for Tannins

Five ml of the extract and a few drops of 1 per cent lead acetate were added. A yellow precipitate was formed, indicates the presence of tannins.

2.3.7. Test for Phenolic Compounds

To 2 ml of filtered solution of the aqueous macerate of the plant material, 3 drops of a freshly prepared mixture of 1 ml of 1 per cent ferric chloride and 1 ml of potassium ferrocyanide was added to detect phenolic compounds. Formation of bluish-green colour was taken as positive.

3. Results

3.1. Phytochemical Screening

Ethanol and aqueous extracts of the selected traditional medicinal plant were subjected to qualitative phytochemical analyses (Tables 13.1–13.5). When performed qualitative tests for phytochemicals in *Vitex negundo, Ocimum tenuiflorum, Ocimum gratissimum Melia azadirachta, Tamarindus indica,* a number of phytochemicals shows positive results in their specific tests.

In the present study, the phytochemical screening of five medicinal plants namely *Vitex negundo, Ocimum tenuiflorum, Ocimum gratissimum Melia azadirachta, Tamarindus indica* showed positive results for saponins, tannins flavonoids, glycosides, alkaloids, carbohydrates, steroids and protein

Extracts of *V.negundo* answered positively alkaloids, saponins, steroids, cardiac glycosides and gum (Table 13.1; Plate 13.2).

In *O. tenuiflorum*, showed the phytochemicals like alkaloids, saponins, tannins, cardiac glycosides and gum (Table 13.2; Plate 13.2).

Extracts of *O. gratissimum* answered positively alkaloids, tannins, cardiac glycosides and gum (Table 13.3; Plate 13.2).

In *M. azadirachta*, showed the phytochemicals alkaloids, flavonoids, steroids, tannins and gum (Table 13.4; Plate 13.3).

Extracts of *T.indica* answered positively alkaloids, flavonoids, phenolic compounds, cardiac glycosides and fixed oil (Table 13.5; Plate 13.3).

Table 13.1: Qualitative Phytochemical Analysis of *Vitex negundo* L.

Phytochemicals	*Ethanol*	*Water*
Alkaloids	+	–
Flavonoids	–	–
Saponins	–	+
Steroids	+	–
Tannins	–	+
Phenolic compounds	–	–
Cardiac glycosides	+	–
Gum	–	+
Fixed fat oil	+	+
'+' indicates: Present		'–' indicates: absent

'+' indicates: Present; '–' indicates: absent.

Table 13.2: Qualitative Phytochemical Analysis of *Ocimum tenuiflorum* L.

Phytochemicals	*Ethanol*	*Water*
Alkaloids	+	–
Flavonoids	+	–
Saponins	–	+
Steroids	–	–
Tannins	+	–
Phenolic compounds	–	–
Cardiac glycosides	+	–
Gum	+	–
Fixed fat oil	–	+

'+' indicates: Present; '–' indicates: absent.

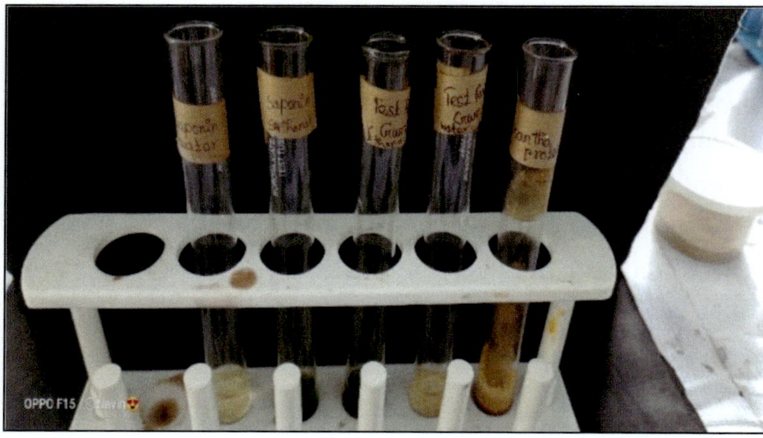

***Vitex negundo* L.**

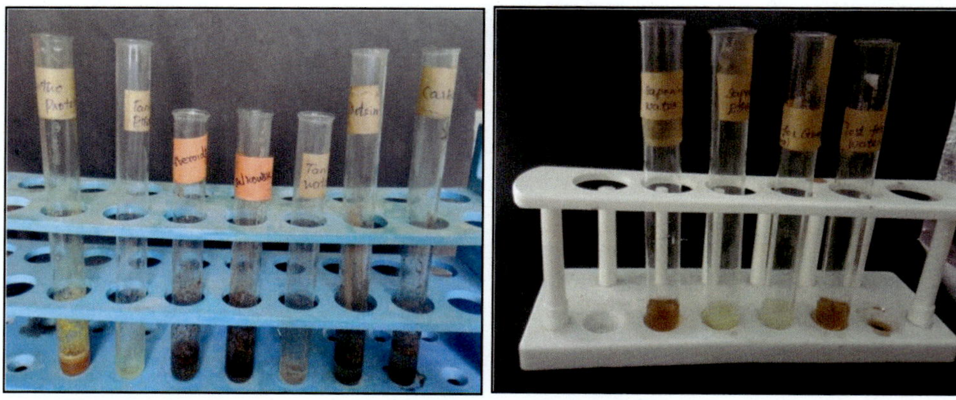

***Ocimum tenuiflorum* L.**

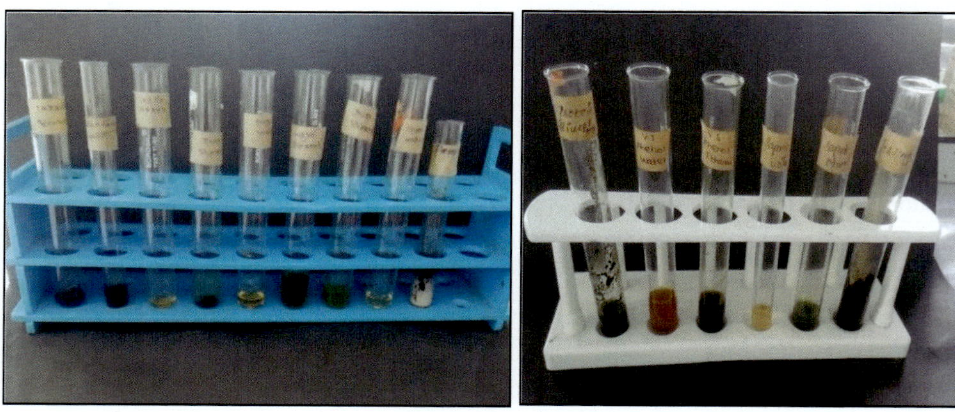

***Ocimum gratissimum* L.**

Plate 13.2: Phytochemical Analysis of Traditional Medicinal Plants.

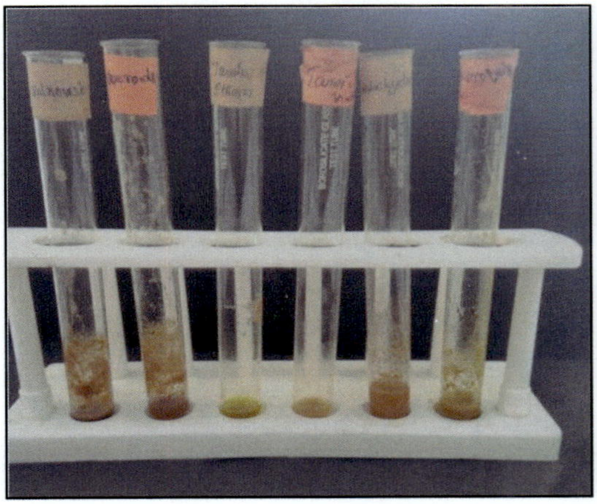

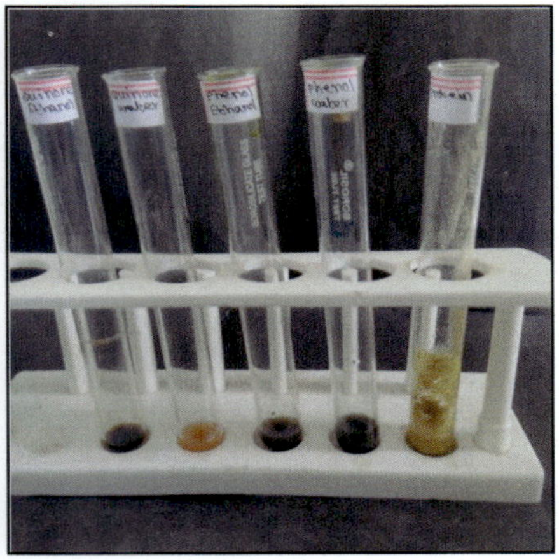

Melia azadirachta **L.**

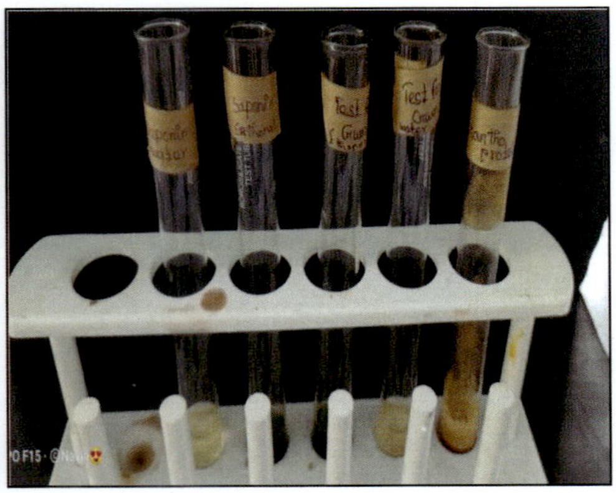

Tamarindus indica **L.**

Plate 13.3: Phytochemical Analysis of Traditional Medicinal Plants.

Table 13.3: Qualitative Phytochemical Analysis of *Ocimum gratissimum* L.

Phytochemicals	*Ethanol*	*Water*
Alkaloids	+	–
Flavonoids	–	–
Saponins	–	–
Steroids	–	–
Tannins	+	+
Phenolic compounds	–	–
Cardiac glycosides	+	–
Gum	+	+
Fixed fat oil	+	+

'+' indicates: Present; '–' indicates: absent.

Table 13.4: Qualitative Phytochemical Analysis of *Melia azadirachta* L.

Phytochemicals	*Ethanol*	*Water*
Alkaloids	+	–
Flavonoids	+	–
Saponins	–	+
Steroids	+	–
Tannins	+	+
Phenolic compounds	–	–
Cardiac glycosides	–	–
Gum	+	+
Fixed fat oil	+	+

'+' indicates: Present; '–' indicates: absent.

Table 13.5: Qualitative Phytochemical Analysis of *Tamarindus indica* L.

Phytochemicals	*Ethanol*	*Water*
Alkaloids	+	–
Flavonoids	+	–
Saponins	–	–
Steroids	–	–
Tannins	–	–
Phenolic compounds	+	–
Cardiac glycosides	+	–
Gum	–	–
Fixed fat oil	+	+

'+' indicates: Present; '–' indicates: absent.

4. Discussion

Secondary metabolites are the classes of compounds which are known to show curative activity against several human ailments and therefore could explain the use of traditional medicinal plant for treatment of some illnesses. In conclusion, the study findings support the use of these five plants in the treatment of infectious diseases caused by resistant microorganisms. These plants can also be used to discover bioactive natural products that may serve as leads for the development of new pharmaceuticals.

All plant parts synthesize some chemicals in themselves which metabolize their physiological activities. In our present study, the selected plants have exhibited different kinds of secondary metabolites. The medicinal value of these secondary metabolites is due to the presence of chemical substances that produce a definite physiological action on the human body. The most important of these include: alkaloids, glucosides, steroids, flavonoids, fatty oils, resins, mucilages, tannins, gums, phosphorus and calcium for cell growth, replacement, and body building (Kubmarawa *et al.,* 2008). A variety of herbs and herbal extracts contain different phytochemicals with biological activity that can be of valuable therapeutic index. It is due to the presence of non-nutritive plant phytochemicals. Different phytochemicals have been found to possess a wide range of activities, which may help in protection against chronic diseases. For example, the phenolics such as flavonoids, tannins are the group of compounds that act as primary antioxidant or free radical scavengers.

Alkaloids have been well investigated for many pharmacological properties including antiprotozoal, cytotoxic, antidiabetic (Cherian and Augusti, 1995) and anti-inflammatory (Liu, 2003) properties but there are only few reports about their antimicrobial properties. The phenolics such as flavonoids, tannins are the group of compounds that act as primary antioxidant or free radical scavengers. Flavonoids are referred as natures biological response modifiers, because of their inherent ability to modify the body's reaction to allergies, anti-inflammatory, antimicrobial, anticancer activities. Apart from the flavonoids shows potent antihyperglycemic activity in animal studies.

Tannins were reported to exhibit antidiabetic (Cherian and Augusti, 1995), anti-inflammatory (Latha *et al.,* 1998), antibacterial and antitumor activities. It has also been reported that certain tannins were able to inhibit HIV replication selectively besides use as diuretics. Plant tannins have been widely recognized for their pharmacological properties and are known to make trees and shrubs a different meal for many caterpillars (Haslem, 1989). Saponins are glycosides occurring widely in plants. They are abundant in many foods consumed by animals and man. Saponin is used as mild detergents and in intracellular histochemstry staining to allow antibody access to intracellular proteins. In medicine, it is used in hypercholesterolemia, hyperglycemia (Rupasinghe *et al.,* 2003), antioxidant, anti-cancer, anti-inflammatory (Manach *et al.,* 1996), central nervous system activities (Argal and Pathak, 2006) and weight loss *etc.* It is also known to have antifungal properties (Rupasinghe *et al.,* 2003).

Plant steroids are known to be important for their cardiotonic activities, possession of insecticidal, anti-inflammatory (Akindele and Adeyemi, 2007), analgesic properties (Malairajan *et al.,* 2006), central nervous system activities (Argal and Pathak, 2006) and antimicrobial properties. They are also used in nutrition, herbal medicine and cosmetics. Glycosides were reported to exhibit anti diabetic characteristics (Cherian and Augusti, 1995). Besides the above said phytochemicals terpenes were found to stimulatesecretion or possess insulin - like effect and decrease blood sugar level in animal studies. Similarly, terpenoids and vitamins acts as regulators of metabolism and play a protective role as antioxidants. Steroids and triterpenoids showed the analgesic properties (Malairajan *et al.,* 2006).

Cardiac glycosides on the other hand are known to hamper the Na +/K + pump. This results in an increase in the level of sodium ions in the myocytes which then enhance in the level of calcium ions. This consequently increases the amount of Ca 2+ ions available for contraction of the heart muscle, which improves cardiac output and reduces distention of heart and thus are used in the treatment of congestive heart failure and cardiac arrhythmia.

5. Conclusion

The plant extracts showed the presence of some secondary metabolites such as alkaloids, flavonoids, steroids, terpenoids, glycosides, saponins and phenolic compounds which were determined by the preliminary qualitative phytochemical analysis. The plant material was extracted with distilled water (aqueous) and ethanol. The preliminary screening of the ethanol extract showed more potent than the aqueous extracts. Thus, the present study, concluded that there is a need of further advancement in the pharmacological investigations of the plant in the field of clinical study of cancer is essential and encourageable.

6. References

1. Akara, E. U., Emmanuel, O., Ude, V. C., Uche-Ikonne, C., Eke, G., and Ugbogu, E. A. (2021). Ocimum gratissimum leaf extract ameliorates phenylhydrazine-induced anaemia and toxicity in Wistar rats. Drug metabolism and personalized therapy, 36(4), 311-320.
2. Akindele, A.J. and Adeyemi, O.O. 2007. Anti-inflammatory activity of the aqueous leaf extracts of *Byrsocarpus coccineus. Fitoterapia* 78: 25 – 28.
3. Argal, A. and Pathak, A.K. 2006. CNS activity of *Calotropis gigantea* roots. *J. Ethnopharmacol.* 106: 142-145.
4. Chaiyana, W., Anuchapreeda, S., Punyoyai, C., Neimkhum, W., Lee, K., Lin, W., Lue, S., Viernstein, H., and Mueller, M. (2018). Ocimum sanctum Linn. as a natural source of skin anti-ageing compounds. Industrial Crops and Products, 127, 217–224. https: //doi.org/10.1016/j.ind crop.2018.10.081
5. Cherian, S. and Augusti, K.T. 1995. Insulin sparing action of leucopelargonidin derivative isolated from *Ficus benghalesis* Linn. *Indian J. Exp. Biol.* 33: 608- 611.
6. Cohen, M. M. (2014). Tulsi-Ocimum sanctum: A herb for all reasons. Journal of Ayurveda and integrative medicine, 5(4), 251.
7. De Caluwé, E., Halamouá, K., and Van Damme, P. (2010). Tamarindus indica L.–A review of traditional uses, phytochemistry and pharmacology. Afrika focus, 23(1), 53-83.
8. Dharmani, P., and Palit, G. (2006). Exploring Indian medicinal plants for antiulcer activity. Indian journal of pharmacology, 38(2), 95-99.
9. Dharmani, P., Kuchibhotla, V. K., Maurya, R., Srivastava, S., Sharma, S., and Palit, G. (2004). Evaluation of anti-ulcerogenic and ulcer-healing properties of Ocimum sanctum Linn. Journal of ethnopharmacology, 93(2-3), 197-206.

10. Ghangale, G. R., Tushar, M., and Jadhav, N. D. (2009). Evaluation of antiulcer activity of Ocimum sanctum in rats. Veterinary world, 2(12), 465-466.
11. Haslem, E. 1989. Plant polyphenols: vegetable tannins revisited-chemistry and pharmacology of natural products. Cambridge University Press. Pp. 169 (Original not seen).
12. Kareti, S. R.; Rajpoot, V. S. Plant Biosystems-An International Journal Dealing with all Aspects of Plant Biology. 2022, 156, 1019-1038
13. Kubmarawa, D., Khan, M.E., Punah, A.M. and Hassan. 2008. Phytochemical screening and antibacterial activity of extracts from *Parkia clappertoniana* keay against human pathogenic bacteria. *J. Med. Pl. Res.* 2(12): 352-355.
14. Kumari, P., Kumari, C., and Singh, P. S. (2017). Phytochemical screening of selected medicinal plants for secondary metabolites. Int. J. Life. Sci. Scienti. Res, 3(4), 1151-1157.
15. Latha, R.M., Geetha, T. and Varalakshmi, P. 1998. Effect of *Vernonia cinerea* Less flower extract in adjuvant induced arthritis. *Gen. Pharmacol.* 31: 601-606.
16. Liu, R.H. 2003. Health benefits of fruit and vegetables are from additive and synergistic combinations of phytochemicals. *Am. J. Clin. Nutr.* 78: 517 - 520.
17. Malairajan, P., Geetha, G., Narasimhan, S. and Jessi Kala Veni, K. 2006. Analgesic activity of some Indian medicinal plants. *J. Ethnopharmacol.* 19: 425-428.
18. Manach, C., Regerat, F. and Texier, O. 1996. Bioavailability, metabolism and physiological impact of 4-oxo-flavonoids. *Nutr. Res.* 16: 517-544.
19. Mapfumari, S.; Nogbou, N.; Musyoki, A.; Gololo, S.; Mothibe, M.; Bassey, K. Phytochemical Screening, Antioxidant and Antibacterial Properties of Extracts of Viscum continuum E. Mey. Ex Sprague, a South African Mistletoe Sipho. Plants 2022, 11, 2094.
20. Marquardt, P., Seide, R., Vissiennon, C., Schubert, A., Birkemeyer, C., Ahyi, V., and Fester, K. (2020). Phytochemical characterization and in vitro anti-inflammatory, antioxidant and antimicrobial activity of Combretum collinum Fresen leaves extracts from Benin. Molecules, 25(2), 288.
21. Meena, A. K., Niranjan, U. S., Yadav, A. K., Singh, B., Nagaria, A. K., Gaurav, A., and Rao, M. M. 2010. Vitex negundo Linn; a Review on Its Ethnobotany, Phytochemical and Pharmacological Profile. Research Journal of Pharmacognosy and Phytochemistry, 2(2), 122-128.
22. Pandey, A. K., Singh, P., and Tripathi, N. N. (2014). Chemistry and bioactivities of essential oils of some Ocimum species: an overview. Asian Pacific Journal of Tropical Biomedicine, 4(9), 682-694.
23. Ramya, S, PJ Jepachanderamohan, N Alaguchamy, M Kalayanasundaram and R Jayakumararaj, 2009. In Vitro Antibacterial Prospective of Crude Leaf Extracts of Melia azedarach Linn. against Selected Bacterial Strains. Ethnobotanical Leaflets, 13: 254-58.

24. Rao A, Kumari S, Laura JS., Dhania. 2023. Qualitative phytochemical screening of medicinal plants using different solvent extracts. Orient Jounal of Chemistry. 39 (3).

25. Rupasinghe, H.P., Jackson, C.J., Poysa, V., di Berado, C., Bewley, J.D. and Jenkinson, J. 2003. Soyasapogenol A and B distribution in Soybean (*Glycine max* (L.) Merr.) in relation to seed physiology, genetic variability and growing location. *J. Agric. Food Chem.* 51: 5888-5894.

26. Shibula, K.; Velavan, S. Determination of phytocomponents in methanolic extract of Annona muricata leaf Using GC-MS technique. Int. J. Pharmacogn. Phytochem. Res. 2015, 7, 1251-1255

27. Singh, D., and Chaudhuri, P. K. (2018). A review on phytochemical and pharmacological properties of Holy basil (Ocimum sanctum L.). Industrial Crops and Products, 118, 367-382.

28. Singh, S., and Majumdar, D. K. (1999). Evaluation of the gastric antiulcer activity of fixed oil of Ocimum sanctum (Holy Basil). Journal of ethnopharmacology, 65(1), 13-19.

29. Singh, S., and Majumdar, D. K. (1999). Evaluation of the gastric antiulcer activity of fixed oil of Ocimum sanctum (Holy Basil). Journal of ethnopharmacology, 65(1), 13-19.

30. Tanko, Y., Magaji, G. M., Yerima, M., Magaji, R. A., and Mohammed, A. (2008). Anti-nociceptive and anti-inflammatory activities of aqueous leaves extract of Ocimum Gratissimum (Labiate) in Rodents. African Journal of Traditional, Complementary and Alternative Medicines, 5(2), 141-146.

31. Trease, G. E. and Evans, W. C. 1997. *Pharmacognosy*, Ed. XIV, Mac Millian Publisher Ltd., London. pp. 568 – 578.

32. Zakir-ur-Rahman, Shahnaz Ahmad, Shamim Qureshi, Atiq-ur-Ranman and Yasmeen Badar, 1991. Toxicological studies of Melia azedarach L. (flowers and berries). Pakistan J. Pharm. Sci. 4(2): 153-158.

Index

A

B

C

R

S

T

U

V

W